Ethnomedicines of Chothe Tribe, Manipur

THE AUTHOR

Dr. Cheithou Charles Yuhlung completed his Graduation (2000) from St. Edmund's College, Shillong, M.A. (2002) and was awarded Ph.D degree (2011) from the Dept. of Sociology, North Eastern Hill University (NEHU), Shillong, Meghalaya. Completed Post-Doctoral Research (2018) from the Dept. of Anthropology, Guahati University, Guwahati, Assam. He has worked as Research Assistant (2006) in an evaluation team for an International NGO (IWGIA) and as Research Associate (2011-12) in an UPE-Project, Dept. of Philosophy, NEHU, Shillong. He has served as Assistant Professor (Sociology) 2012-14 in Women's College, Shillong and also taught Sociology in William Carey University, Shillong, Meghalaya. He has published several original research papers in reputed journals like *Sociological Bulletin* (2007), *IOSR-JHSS, Tribal Bulletin, Asian Man and in others*. He served as the General Secretary of Northeast India Tribal Research Association (NEITRA) and had edited a book entitled as *'Northeast India Tribal Studies: An Insider's View'*. He has published books entitled as *'Chothe Indigenous Religions and Culture: A Sociological Study'* 2018 & *Ethnomedicinal Plants of Maring Tribe of Manipur: NE India*. He has participated in various International and National seminars and conferences.

Ethnomedicines of Chothe Tribe, Manipur

Cheithou Charles Yuhlung

2021

Regency Publications®

A Division of

Astral International Pvt. Ltd.

New Delhi – 110 002

Cataloging in Publication Data--DK
Courtesy: D.K. Agencies (P) Ltd. <docinfo@dkagencies.com>

Yuhlung, Cheithou Charles, author.
Ethnomedicines of Chothe tribe, Manipur / Cheithou Charles Yuhlung.
pages cm
Study conducted in Bishnupur, Chandel and Tengnoupal districts of Manipur, India.
Includes bibliographical references and index.
ISBN 9789354614552 (HB)

1. Chothe (Indic people)--Medicine--India--Bishnupur (District) 2. Chothe (Indic people)--Medicine--India--Chandel (District) 3. Chothe (Indic people)--Medicine--India--Tengnoupal (District) 4. Medicinal plants--India--Bishnupur (District) 5. Medicinal plants--India--Chandel (District) 6. Medicinal plants--India--Tengnoupal (District) I. Title.

LCC GR880.Y84 2021 | DDC 615.88095417 23

Published by : **Regency Publications®**
A Division of
Astral International Pvt. Ltd.
– ISO 9001:2015 Certified Company –
4736/23, Ansari Road, Darya Ganj,
New Delhi-110 002
Ph. 011-43549197, 23278134
E-mail: info@astralint.com
Website: www.astralint.com

Acknowledgement

In the first place, I thanked the Almighty God for all his love and blessings bestowed upon me in this endeavour to help complete my *Post-Doctoral* higher research work and enabling me to publish this second book under the study. I also expressed my sincere gratitude to my family members and wife Elizabeth for their constant prayers, support and encouragement.

I am deeply honoured to Prof. Mini Bhattacharyya Thakur (former HOD), Dept. of Anthropology, Gauhati University, Assam who happily consent to be my mentor and supervisor when I initially approached her. Since then, she has been providing me invaluable suggestions and guidance, and she was always there for me whenever I needed her help. I also expressed my gratitude and thanked to Prof. Rekha Das (Retd.) and Prof. AA Ashraf (HOD), for their encouragement and support. I also extend my gratitude to all the faculty members, office staffs and fellow research scholars of the Department of Anthropology, Guahati University, Assam for their love, support and cooperation in the department.

The author sincerely thanked the *University Grant Commission* (UGC), New Delhi for awarding him the *Post-Doctorial Fellowship* under the scheme *for the Scheduled Caste and Tribe* (SC/ST), awarded in 2012 for the financial assistance to pursue this research.

The author expressed his indebtedness and thanked the Chothe respondents like H. Thambaljao, Y. Tomalsing, Pr. Roushi, M. Wounchung, Kh, Khedon, Mrs. Thambalhoi, M. Merachou, M. Kundo, Mrs. Y. Atonsicha, Mrs. Nalini, Mrs. Poinuleima, Mrs. H. Atonpi, Th. Agollu, Mrs. H. Hoinu, Mrs. Th. Yumchaoleima, and M. Cingsianniang (Niangboi) and many others who are the village chiefs, village elders, senior citizens, local medicine man, priest, priestess and leaders. This Post-Doctoral research would not have been successfully completed without their kind cooperation and invaluable information provided herein.

I also expressed my sincere appreciation and thanked to certain friends like Mr. Heshu Aji Mao, John Kayina Mao, Dr. Jerry, Saleo, Jytishman, Arun, Nancy, Humi Thaosen, Chintu, Shangpam, and other friends who are always there for me and kept company during my higher research work and stayed in Guwahati, Assam. Lastly, but not the least, I expressed my sincere thanked you to all those persons whose names are not mentioned herein but have helped me in different ways during my work. Once again, in a special way, I thanked each and everyone involved in this work and has had helped me in completion of this higher research work.

Thank You and God Bless You All

Dr. Cheithou Charles Yuhlung

(Post-Doctoral Fellow)
Dept. of Anthropology,
Gauhati University,
Guwahati, Assam – 780014

Preface

The book entitled *'Ethnomedicines of Chothe Tribe, Manipur'* is based on the research work among the indigenous Chothe tribe of Manipur under the UGC Post-Doctoral Fellowship. The study covers the Chothe villages located in three districts *viz.*, Bishnupur, Chandel and Tengnoupal. The data was collected using questionnaires, interviews and interview-scheduled based on a random sampling technique among the select village elders, priest, priestess, village leaders and senior citizens of Chothe. The information was collected by participant observation method.

The book is divided into five chapters. The Chapters One - introduces the significance of the book. Chapter Two - describes the identified ethno-botanical plants used as Ethnomedicines by the Chothes. Chapter Three – briefly elaborates on ethnobotanical plants that have religious significances besides, other forms of indigenous medicines practiced by the Chothes. Chapter Four – deals with data analysis and interpretation presented in table forms. Chapter Five – is the conclusion.

In the study, there are a total of 130 (one hundred and thirty) identified ethno-botanical plants which belongs to 58 families that are traditionally used as ethnomedicines by the Chothes of Manipur to treat about 69 different ailments and diseases. The most common families with maximum frequencies are Lamiaceae, Zingiberaceae, Fabaceae and Asteraceae.

And certain vulnerable diseases are like fever, cough, dysentery to diabetes, jaundice, cancer, urinary/kidney stone cases, irregular menstrual problems besides other ailments.

The book also discusses certain ethno-botanical plants that have religious significances in the practice of healing and curing ailments that are often used to protect and ward off evil spirits from falling trap to their wraths. Besides, it also briefly highlights how the indigenous Chothe practice certain animal's part, insects and other objects in healing and curing different types of sicknesses. All these are also presented in Table. There are nine tables in all.

Despite the development of pharmaceutical medicines, the Chothe still used their ethnomedicines side by sides in different ways to cure and heal certain ailments and diseases. Most of the tribes of Manipur lack the knowledge of herbarium since, the plants are grown wild on the roadsides, gardens and in the interior forest. Some highly valuable seasonal plants are dried and kept safely at home for future used.

However, there has been sharp declined in the practice of ethnomedicines among the Chothes and other tribes of Manipur with the declined of knowledgeable elderly persons in the villages and with the reluctancy of the younger generations to abide the old traditional practices since, they arc already influenced with the modern living styles. Therefore, certain serious steps should be taken up to encourage the people to document, revive and continue their indigenous ethnomedicinal practices in a scientific manner. The study is basically a documentation to understand how seriously and effectively the Chothes still continue to practice their ethnomedicines in the contemporary world.

Dr. Cheithou Charles Yuhlung

Contents

Abbreviations

H: Herb

S: Shrub

T: Tree

C: Creeper

Cl: Climber

S: Seasonal

P: Perennial

O: Oral

E: External

List of Tables

Chapter 1

Introduction

The indigenous healing and curing system using various ethno-botanical plants, animal's parts, insects and other objects among different tribes and communities had been practising since early days. For example, when a person gets cut or wounded and bleeds excessively, he would look out for certain herb or bark of a tree, and in which, he would crush or chewed it and apply the paste immediately on the wound to help the blood coagulate or stop bleeding. Such traditional treatment method of healing and curing is still practice even today by many tribal communities of the world like Chothe, Maring, Chiru and other Southeast Asian tribes despite, the advancement in pharmaceutical medicines. The book listed 130 identified ethno-botanical plants belonging to 58 families that treat more than 69 ailments and diseases. It also provides other indigenous ethnomedicines like animal's part, insect, bird, and other substances used in curing certain ailments and sicknesses by the Chothes. It indicates that certain ethno-botanical plants are consume in different forms as local chutney or simple boiled curry thereby providing them rich nutrients and minerals as food supplements that are highly required in the body systems.

The Chothe Tribe

The Chothe is a minor indigenous tribe of Manipur, situated in the North-Eastern region of India. They are spread in three districts of Manipur, *viz.*: Bishnupur, Tengnoupal and Chandel with eighteen (18) villages (*i.e.* twelves census recognized villages and six new hamlets). The 2011 Census of India provides the total Chothe population as 3585 (1706/M-1879/F) with a literacy rate of 69.79 %[1].

Historically, the Chothe is described as one of the oldest tribes that migrated to southern Manipur much before other new ethnic tribes entered[2,3,4]. They have distinctive language, kinship, culture, religion, economy and political institution. They belong to the Mongoloid racial stock and speaks Sino-Tibetan language of Tibeto-Burman family under the Chin-Kuki and Naga-Kuki linguistic group as 'Old-Kuki' speakers[5].

The Chothe is a recognized *Scheduled Tribe* of Manipur under the Government of India Act of 29th October 1956[6]. The majority of Chothes are now Christians, and only a handful of them especially, from Lamlanghupi village of Bishnupur district still adheres to their indigenous faith who upholds their traditions and customs. The various neighbouring tribes of Chothes are like Aimol, Chiru, Kom, Anal, Moyon, Monsang, Lamkang, Tarao, Maring, Rongmei (Kabui/Zeliangrong), Thadou speaking groups (New-Kuki) and the Meitei.

Statement

The Chothes inhabit in the state of Manipur, Northeast India. The region is also well-known hotspot zone of 'Cultural-diversity' since, the region is inhabited by about 220 distinctive tribes, sub-tribes and ethnic groups having its own distinctive cultural traits[7]. Manipur has 33 recognised scheduled tribes besides, OBC, castes and Meitei. Some tribes of Manipur though share similarity with Chothes in many respects yet, certain aspects are completely different from one another in custom, tradition and village administration.

India is one of the 12 mega-biodiversity countries of the world with 45,000 plant species while the concentrated hotspots of India are the North-Eastern region, Western Ghats and Andaman and Nicobar Islands[8]. Jaiswal said till date about 3,000 plants have been officially documented as having medicinal potential, but more than 6,000 plants are used by the traditional herbal practitioners[9].

This North-eastern region of India is very rich with diverse flora, fauna and several species. The region is identified and ranks 8th amongst the 34 'Bio-diversity Hotspots' region of the world[10, 11]. It is situated in temperate tropical rain forests zone within 23°51′ N and 25°41′ N Latitudes and 93°3′ E and 94°4′ E longitude bordering Myanmar[12].

The Chothe, like many indigenous tribes of the world do practice ethno-medicines when a person suffers from certain ailments, sicknesses and diseases. This traditional method of healing and curing has been handed down orally from one generation to the next in the form of folklore, folktales, and sayings. Thus, we know that in the past, before the development of advanced scientific technology and pharmaceutical medicines almost every community big or small, rich or poor used various ethno-botanical plants and other substances as medicines in the treatment of various humans and animal's ailments, sicknesses and diseases.

The contradiction observed today is that the urban dwellers depends on advance pharmaceutical medicines that are available in their cities and towns, while majority of the rural people still depends on various ethno-botanical plant's parts (trees, shrubs, herbs and rhizomes) and other substances available in their gardens and surrounding environment. Both are considered effective in many ways in fighting against any virus, germs and toxic poisonous substances. Although, at one point of time, people assumed that advance pharmaceutical medicines were the best in curing and healing different sicknesses and diseases, but later, the reports resulted with many side-effects problems. Thereby, modern man re-researches on the practices of various ethnomedicines of the folks and found that the reports have no or less side-effects remarks. Therefore, Governments have taken up steps to encourage in the development of Ayurveda, Homeopathy, Allopathy, and Herbal medicines for the remedy of different ailments and diseases.

However, with globalization, development in science and technology, advance in pharmaceutical medicines, influences of modernization, westernization, proselytization to Christianity, deforestation, disturbance in environments, climate change, and gradual decline in the number of traditionally knowledgeable individuals especially, among the tribal communities have raise certain major problems in the process of collecting the required information on the practices of ethnomedicines among the tribal communities because they did not maintain a written record. In the past, the valuable information was handed down orally and traditionally from one generation to another and to their most trustees individual or followers. But with the passage of time, majority of the traditional know-how elders have died without leaving complete information to the younger generations, and hence, there have been vast knowledge gaps in the practice of indigenous ethnomedicines.

The reasons of negligence are many. Most important aspect is that the younger generations are reluctant to accept and follow their indigenous traditional treatment methods with a belief that it involves magico-religion, superstition and is in-effective. In the context of the past, these beliefs are also true to certain extend because some priests/priestess involves the practiced of magico-religion and animal's sacrifices in healing and curing certain illnesses. Moreover, the younger generations are already influenced by modern medicines and western living styles. Third reason is that, some village priests kept their expertise and talent top secret and confides only to their most trustees. Fourthly, deforestation and climate changed is also another major factor since it has disturbed the environment and ecology causing certain plants from natural growth and threatens their survival and becomes endangered species now. Fifth, jhuming or shifting cultivation and wild fires have made certain valuable medicinal plants extremely rare and scarcity. Sixth, over exploitation of natural resources from the natural habitat like Ginsen and black turmeric are also some examples. The practice of setting up herbarium to preserve their valuable lifesaving drug plants among the tribals is very much ignore and neglected. One can say that there is none among the tribals of Manipur, although some Meiteis do maintain such small herbarium.

The above factors of various changes, death, impacts, global developments, deforestation, environment degradation, people's perception and attitudes have posed a serious threat to the rich traditional practices of ethnomedicines among different ethnic groups and tribes of the world. Therefore, in this context, a study primarily focused in documentation and understanding the extensiveness and its relevancies in the use of various ethno-botanical plants as ethnomedicines among the Chothes of Manipur was carried out before it completely disappeared. The outcome of the research work is provided herein.

Objective of the Study

The primary objective of the study was to explore and document the practices of ethno-medicines among the Chothes in the treatment of various ailments, sicknesses and diseases since, no serious study has been carried out in this area. The rapid declined in usage, ignorance, climate change, environment degradation and advancement in pharmaceutical medicines has posed a serious threat to the decline of the indigenous tradition of ethnomedicines. The study is expected to have immense value and relevance for the future generations, as well as enriched the existing knowledge in the field of indigenous ethno-medicinal studies especially among students, research scholars and certain laymen.

Area of the Study

The study area is based in the state of Manipur, India. The study covers the indigenous Chothes community inhabiting in three districts of Bishnupur, Tengnoupal and Chandel.

Materials and Methods

The study was carried out among 40 select Chothe respondents (20 males and 20 females); represented by village chiefs, priests, village elders and senior citizens. Besides, these 40 respondents there are many other important personalities who provided valuable information about the society and helped in this research survey. The primary sources of data are collected using interview, interview-scheduled guide techniques as its tool by going to the field. For the comparative reference analysis of the data, secondary sources like published articles, books and materials available in different websites are also consulted.

Literature Reviews

In the Northeast region of India, there have been various scientific undertakings of exploration, identification and mapping of various flora and fauna and to explain the unexplained phenomena about the people, culture and area from different context.

Chakraborty said that the North-Eastern ethnic communities based their traditional knowledge of medicine on their needs, instinct, observation, trial and error, and long experience in the healing methods. Such knowledge often served as an important part of their cultural identities[13].

Pfoze (2012) said that it is estimated that a total of 60% of the world population and 80% of the population in developing countries depend on traditional medicines, mostly plant drugs for their primary health care needs[14]. In most developing countries like India the flora remain virtually unexplored from practical utilization, yet past experiences have shown that many valuable drugs have been derived from plants[15].

Singh and Arora (1978) states that about 800 species of wild edible plants occur in different floristic regions are consumed by the tribal communities of the region, and that about 50 % of the plants identified from the region have medicinal values[16]. Mao AA stated that about 50% of the total 17,500 flowering plants hail from the region, and 40% of them are endemic[17].

Therefore, the *World Health Organization* (WHO) recognized the healing properties of these traditional medicinal herbs and thus suggests research should be stepped up[18, 19]. Hazarika said that WHO estimates about 80% of the population of most developing countries relies on herbal medicines for their primary healthcare need[20].

Shankar (2013) said medicinal plants exploration in North Eastern India has been carried out in the various forests including crude drugs markets up to the range wherever medicinal plants produce are sold in the markets[21]. Various attempts have also been made for the acclimatization through cultivation of medicinal plants from one zone to another in the Region. The crude drugs produce from the region are traded in the markets of Assam, West Bengal, Bihar and even in the Central market of Delhi. Despite the news of over exploitation of medicinal plants in the region, alertness has created among the dedicated researchers to continue their exploration and document the various ethno-medicines amongst various ethnic tribes and groups.

Thus, many local Indian scholars have begun their ethno-medicinal studies among various North-Eastern tribes. Some of them are like Shankar-Mishing[22], Khongsai-Arunachal Pradesh[23]; Ripunjoy- Kachari[24], Das- Cachar[25], and Basumatary-Bodo of Assam[26]; Jamir- Ao of Nagaland[27]; Singh- Khasi[28], and Jaiswal- Jaintia of Meghalaya[29]; Rajkumari- Chiru[30], and Pfoze- Mao[31], Prakash- Rongmei[32], Yumnam-Meitei[33], Singh[34], Jain- of Manipur[35]; Rai- Mizoram[36], and Yuhlung – Maring of Manipur[37] and so on.

Endnotes

[1.] Census of India 2011. (Population census).

[2] Shakespeare, J. 1912. The Lushai-Kuki clans. Suman Lata, Cultural Publishing House, Delhi.

[3] Ansari, SA. 1991. Manipur: Tribal demography and Socio-economic Development. Delhi: Anil Mittal.

[4] Roy, R. C. 1936. 'Notes on the Chawte Kuki Clan', in Man in India. Vol.16, pp.135-155.

[5] Greirson, George Abraham (Ed.). 1967. Linguistic Survey of India. Vol.III. Tibetan-Burman Family, Part-III (Specimens of the Kuki-Chin and Burma groups). Calcutta: Superintended of Government Printing), [reprinted, Delhi: Motilal Banarsidass].

[6] See above, Ansari, SA. 1991.

[7] Yuhlung, Cheithou Charles and P.G Jangamlung. 2016. Northeast India Tribal Studies: An Insider's View. New Delhi: Astral International Publication.

[8] Jaiswal, Vidhan, Culture and ethnobotany of Jaintia tribal community Meghalaya Northeast India – A mini review, Indian J Traditional Knowledge, 9(1) (2010) 38-44.

[9] Jaiswal, Vidhan, Culture and ethnobotany of Jaintia tribal community Meghalaya Northeast India – A mini review, Indian J Traditional Knowledge, 9(1) (2010) 38-44.

[10] North Eastern Council Vision (2020). 2012. Ministry of Development of North Eastern Region & North Eastern Council. (NER 2020 - NECLook East Policy. pdf). Vol.1. http://www.indianchamber.org/northeast.

[11] Ningombam, DS; Deshworjit Singh and Potsangbam Kumar Singh. 2014. Ethnobotanical Study of Phologacanthus thyrsiformis Nees: A Conserved Medicinal Plant of Manipur, Northeast India', in International Journal of Herbal Medicine. Vol.1(5), pp. 10-14. www.florajournal.com, accessed on 24/04/14.

[12] Singh, B.H; Singh, P.K., Singh, S.S and Elangbam, B. 1996. Indigenous bio-folklores and practices. Its role in biodiversity conservation in Manipur, in Journal of Hill Research. Vol.9(2), pp.359-362.

[13] Chakraborty, Raja, Biplab De, N Devanna, Saikat Sen. 2012. North-East India an Ethnic Storehouse of Unexplored Medicinal Plants, in Scholars Research Library; Journal of Natural Product and Plant Resource. Vol.2 (1), pp.143-152. http://scholarsresearchlibrary.com/archive.html, accessed on 03-04-2014.

[14] Pfoze, Neli Lokho; Yogendra Kumar, Bekington Myrboh. 2012. Survey and assessment of ethnomedicinal plants used in Senapati District of Manipur State, Northeast India, in Phytopharmacology. Vol.2(2), pp.285-311. *www.phytopharmajournal.com*, accessed on 02-09-2014.

[15] Singh, V.K and Abrar M. Khan. 1989. Medicinal plants and folklores: A strategy towards conquest of human ailments. New-Delhi: Today and Tomorrow's Printer & Publishers. (A Glimpses in plant research. Vol.9).

[16] Singh, H.B and Arora, R.K. 1978. Wild edible plants of India, in ICAR. New Delhi.

[17] Mao, AA; TM Hynniewta and M. Sanajaoba. 2009. Plant wealth of Northeast India with reference to ethnobotany, in Indian Journal of Traditional Knowledge. Vol.8(1), Jan. pp. 96-103.

[18] See above, Singh, V.K and Abrar M. Khan. 1989.

[19] Buragohain, Jitu. 2011. Ethnomedicinal Plants Used by the ethnic Communities of Tinsukia District of Assam, India, in Recent Research in Science and Technology. Vol.3(9): 31-42.www.scholarjournals.org, accessed on 03-04-2014.

[20] Hazarika, Ridip; Santoshkumar Singh Abujam and Bijoy Neog. 2012. Ethno Medicinal Studies of Common Plants of Assam and Manipur, inInternational Journal of Pharmaceutical & Biological Archives. Vol.3(4), pp:809-815. www.ijpba.info, accessed on 05-04-2014.

[21] Shankar, Rama and M.S. Rawat. 2013. Conservation and cultivation of threatened and high valued medicinal plants in North East India, in International Journal of Biodiversity and Conservation.Vol. 5(9), September, pp. 584-591. http://www.academicjournals.org/IJBC, accessed on 23/11/13.

[22] Shankar, R; Lavekar GS, Deb S, Sharma BK. 2012. Traditional healing practice and folk medicines used by Mishing community of North East India.J Ayurveda Integr Med. Vol.3(3):124-9. http://www.jaim.in/text.asp

[23] Khongsai; M Saikia, SP and Kayang H. 2014. Ethno-medicinal plants used by different tribes of Arunachal Pradesh, in Indian Journal of Traditional Knowledge. Vol.13(2), April, pp.368-376.

[24] Ripunjoy, Sonowal. 2013. Indigenous Knowledge on the Utilization of Medicinal Plants by the Kachari Tribe of Dibrugarh District in Assam, North-East India. International Research Journal of Biological Sciences. April Vol. 2(4), 44-50. Available online at: www.isca.in

[25] Das, Ajit Kumar; BK Dutta and GD Sharma. 2008. Medicinal plants used by different tribes of Cachar district, Assam, in Indian Journal of Traditional Knowledge. Vol.7(3), July, pp.446-454.

[26] Basumatary, S; M Ahamed and SP Deka. 2004. Some medicinal plant leaves used by Boro (Tribal) people of Goalpara district, Assam, (Explore), in Natural Product Radiance. Vol. 3(2), March-April. http://nopr.niscair.res.in, accessed on 02-03-2014.

[27] Jamir, NS; Lanusunep and Narola Pongener. 2012. Medico-herbal medicine practiced by the Naga tribes in the state of Nagaland (India).Indian Journal of Fundamental and Applied Life Sciences. Vol. 2 (2) April-June, pp.328-333. http://www.cibtech.org/jls.htm,accessed on 02-02-2014.

[28] Singh, K Anurudh. 2010.Probable Agricultural Biodiversity Heritage Sites in India: V. The Garo, Khasi, and Jaintia Hills Region. Asian Agri-History. Vol.14(2), (133-156).

[29] Jaiswal, Vidhan.2010.Culture and ethnobotany of Jaintia tribal community of Meghalaya, Northeast India: A mini review, in Indian Journal of Traditional Knowledge. Vol.9(1), Jan. Pp.38-44.

[30] Rajkumari, Ranjana; P.K. Singh, Ajit Kumar Das and B.K. Dutta. 2013. Ethnobotanical investigation of wild edible and medicinal plants used by the Chiru Tribe of Manipur, India, in Pleione. Vol.7(1), pp:167-174. http://www.pleione.info/, accessed on 02-09-2013.

[31] See above, Pfoze, Neli Lokho.

[32] Prakash, N; MA Ansari, P Punitha and PK Sharma.2014. Indigenous traditional knowledge and usage of folk bio-medicines among Rongmei tribe of Tamenglong district of Manipur, India, in Prakash (et al.), African Journal of Traditional Complementary and Alternative Medicine. Vol.11(3), pp.239-247. http://www.researchgate.net, accessed on 02-09-2013.

[33] Yumnam, Rajesh Singh; CH. Onita Devi, Santosh Kumar Singh Abujamand D. Chetia.2012. Study on the Ethnomedicinal System of Manipur. International Journal of Pharmaceutical & Biological Archive, 3(3):587-591. www.ijpba.info

[34] Singh, Senjam Jinus; V.K Batra, Sanjive Kumar Singh and Thiyam Jefferson Singh.2012. Diversity of underutilized vegetable crops species in North-East India with special reference to Manipur: A review, in NeBIO. Vol. 3(2), June, pp.87-95. www.nebio.in, *accessed on 02-03-2014.*

[35] Jain, Alka; S Roshnibala, PB Kanjilal, RS Singh and H Birkumar Singh. 2007. Aquatic/semi-aquatic plants used in herbal remedies in the wetlands of Manipur, Northeastern India, in Indian Journal of Traditional Knowledge. Vol6(2), April, pp.346-351.

[36] Rai, Prabhat Kumar and H. Lalramnghinglova. 2010. Lesser known ethnomedicinal plants of Mizoram, NorthEast India: An Indo-Burma hotspot region, in Journal of Medicinal Plants Research. Vol.4(13), 4 July, pp. 1301-1307. http://www.academicjournals.org/jmpr, accessed on 02-03-2014.

[37] Yuhlung, Cheithou Charles and Mini Bhattacharyya. 2016. Indigenous medicine used by the Maring tribe of Manipur, Northeast India. Journal of Ayurvedic and Herbal Medicine. 2(4): 146-153.

Chapter 2

Practices of Ethnomedicines among the Chothe of Manipur

Part A (A)

1. ***Acalypha hispida*** **Burm. F**

   Family: Euphorbiaceae

   Common Name: Chenille plant, Red hot cat's tail.

   Local Names – Chothe: *Kaprang arapa;* **Manipuri:** *Kaprang man*

Part Used: Leaf.

Diseases/Treatments: White discharge of woman.

Crush the fresh leaves and add little limestone and apply on the body parts twice a day for a week or till cured.

How to Identify: The shrub plant grows upto a high of about 1-15 feet. The leaves are like mulberry leaves and the flowers are pinkish or red in colour and hairy. The flowers are long about 2-5 inches long.

2. *Acemella oleracea* (L.) R.K. Jansen

Family: Asteraceae

Common Name: Toothache plant, Sichuan buttons

Local Names – Chothe: *Ansa;* **Manipuri:** *Morok sabi*

Part Used: Leaf/Flower.

Diseases/Treatments: i) Boil, ii) Urinary problem, iii) Toothache, (iv) Worms (esp. child).

a. The matured leaf is heated on fire and applied on the boil part, time after time till it is cured.

b. Those suffering from urinary disorder may eat fresh tender leaves as simple boiled curry.

c. When a person suffers from toothache the flower bud may be crushed with hands and inserted in the cavity of the toothache or put around the aching tooth time after time till it gets relief. It acts as sedative.

d. Boil the leaves and give bath with the warm water to the baby 2-3 times occasionally once a month only.

How to identify: This herb grows in a moist wasteland areas upto a high of about 1 ft. The leaves are heart shaped and the flowers are bright yellow with bulgy bud. The species is synonymous with *Spilanthes acmella*. The locals rather term it as rabbit's feeds or grass.

3. *Achyranthes aspera* Linn.

Family: Amaranthaceae

Common Name: Devils horsewhip, Prickly chaff flower, Apang

Local Names – Chothe: *Haotong koite*; **Manipuri:** *Haodong thangol*

Part Used: Leaf.

Diseases/Treatments: Diarrhoea (Esp. for children)

Collect some fresh leaves and cooked it as simple boiled curry along with dried fishes and serve 3 times a day for 3 days. This is especially for children above 2 years. The simple curry should be taken by the feeding mother too daily till cured.

How to Identify: The herb grows wild on roadsides and wasteland areas upto a high of about 1-3 feet. The leaves are small, heart shaped and has long stalk with tiny spiky and hook seed pods. These tiny hooks spiky seed pods when pricked are itchy and slightly painful. The local termed it as the baby's sickness as Khom lai okpa.

4. *Acorus calamus* Linn.

Family: Acoraceae

Common Name: Sweet flag, Beewort.

Local Names – Chothe: *Oak lou;* **Manipuri:** *Oak hidak*

Part Used: Rhizome.

Diseases/Treatments: Worms/Ascaris (esp. for children).

A piece of the rhizome is worn around the child's neck as necklace. Crush little of the rhizome and mixed little pure water and give a tea spoonful (3ml) twice a day only. Do not give excess to children below two years.

How to Identify: It grows wild or cultivated around moist or wetland areas. The leaves are green-yellowish, long and flat like structure soft and tender. The rhizome has fibrous roots around it. The flower is has a long yellow bulb.

5. *Adhatoda vasica* Nees.

Family: Acanthaceae

Common Name: Malabar nut.

Local Names – Chothe: *Annha angoupa;* **Manipuri:** *Nongmangkha Angouba.*

Parts used: Leaf.

Diseases/Treatment: Jaundice

Cook some fresh tender leaves as simple curry along with dried fishes and have twice a day morning and evening for a week. It is strongly suggested to avoid red meat, chilly, fresh fishes). The taste will be little sour.

How to Identify: The bitter shrub grows upto a high of about 5-10 feet. The leaves are elongated, shiny, thick and bitter in taste. The flowers are white and little big. The locals planted it as house fencing. The local term it as white Malabar plant since it belongs to the same family.

6. *Alangium chinense* (Lour.) Harms.

Family: Cornaceae

Common Name: Begonialeaf alangium, Chinese Alangium.

Local Names – Chothe: *Kokal;* **Manipuri:** *Kokal*

Part Used: Leaf.

Diseases/Treatments: Fever

Crush the fresh leaves and take about 10-20ml of the extracted liquid 2 times a day till cured. For children the dosage is given about 2-5ml only. It tastes bitter.

How to Identify: The tree grows upto a high of about 15-20 feet. The leaves are thin and heart shaped but with pointed tips at sideways. The flowers are tiny white and yellow pistil. The fruits are tiny like berries. The flower season is May-June.

7. *Aleo vera* (L.) Burm.f.

Family: Asphodelaceae

Common Name: Aleo vera

Local Names – Chothe: *Kheva arikpa/Aleo vera;* **Manipuri:** *Aleo vera*

Part Used: Leaf.

Diseases/Treatments: i) Burn, ii) Chronic gastritis/ulcer, iii) Dark pigmentation.

a. Immediately apply the *aleo vera* leave gel smoothly over the burn areas till the burning sensation is cold. This will reduce the watering of the skin and burnt marked.
b. Eat the leave gel a spoonful twice a day in empty stomach and drink plenty of water after 15 minutes gap for a week or two weeks to help cure from chronic ulcer and gastritis.
c. To cure the dark pigmentation on the face, apply the leave gel over the dark pigmentation over the face in the morning and evening for a month and also eat a spoonful of the fresh leave gel for two weeks.

How to Identify: There are two common variety found in the region. The plump-spongy is preferred than the thin sleekly one. The flowers are yellow in colour and blooms in bunches from a long stalk.

8. *Allium sativum* Linn.

Family: Alliaceae

Common Name: Garlic

Local Names – Chothe: *Satun;* **Manipuri:** *Chanam*

Part Used: Bulb

Diseases/Treatments: Fever, Rheumatism, Arthritis, Paralysis, Muscular Pains.

When a person suffers from slight fever, rheumatism, arthritis or semi-paralysis may apply the ointment prepared by adding few pods of garlic on the heated some pure mustard oil in a bowl. The fresh garlic bulbs are considered better. This ointment may be applied as balm and also used as massaging ointment in the painful joint areas, forehead, near the nostril, throat cleavage, chest, back, palm and feet till the body temperature returns to normal condition. This massaging if repeated regularly twice a day for a week or more to help reduce fever, rheumatism, arthritis, muscular joint pains.

How to Identify: Besides, serving as daily culinary item the Chothes believed that the smell can dispel the evil spirit. Therefore, when they suffered from bad dreams or nightmares they used to keep it under their pillow or mattress.

9. *Alocasia indica* (Roxb.) Schott.

Family: Araceae

Common Name: Great-leaved Calendium, Giant Taro

Local Names – Chothe: *Jaitim;* **Manipuri:** *Yendem*

Parts used: Stem/Leaf.

Diseases/Treatment: (i) Enhance blood (Post-natal woman), (ii) Boil.

a. Traditionally the stem after removing the fibre or outer covering is cooked in a simple boil curry form along with fresh local fishes (Ngara) and given to post-natal woman to enhance the blood and enhance the stimulation.

b. Smear a piece of the leaf on the palm and add little saliva and apply the leaf on the boil area with little opening on the boil mouth. Repeat it 4-5 times after an interval of 6-8 hours till the pus is mature to burst.

How to Identify: This herb is cultivated for home consumption in the kitchen garden. There are two common varieties green and purple. It grows upto a high of 2-4 feet. The leaves are broad, shiny and heart shaped. Because of its usefulness often the stems are dried in the sun and kept for using during off season. Whether fresh or dried the stems/stalks are occasionally eaten either as chutney or as simple cooked curry with fishes.

10. *Alpinia caerulea* (R.Br.) Benth.

Family: Zingiberaceae

Common Name: Native ginger

Local Names – Chothe: *Phuiphung;* **Manipuri:** *Puleimanbi*

Part Used: Root

Diseases/Treatments: Asthma and Tonsillitis.

Person suffering from asthma and tonsillitis may drink about 10-20ml of the extracted liquid obtained from crushing the roots by adding little salt twice a day till it gets cured the tonsillitis. But for asthma patient one may take about 10ml time to time when required.

How to Identify: This rhizome plant is planted around the house for home consumption. The long soft stalk bears leave after leaves like stem and grows upto a high of about 5 feet. The leaves are elongatedly big, broad and smooth, although abit hard. The flowers blooms in bunches white in colour and bead seeds. The stem or rhizome grows like a creeper underground and new plant grows from the rhizome knots. Take a piece of the leaf and smear it and one will get a strong spicy aromatic smell of the rhizome. The Chothes and other communities of Manipur preserved the dried rhizome to add in beef, pork and fish curry for its unique spicy aroma. This species is synonymous with *Alpinia allughas* (Retz) Roscoe.

11. *Alpinia galangal* Linn. Willd.

Family: Zingiberaceae

Common Name: Greater galangal

Local Names – Chothe: *Marou (Barou)*; **Manipuri:** *Kanghu*

Part Used: Rhizome

Diseases/Treatments: (i) Piles and, (ii) Termination of conception of a child (Abortion).

a. Those persons suffering from piles may drink about 30-40ml of the extracted liquid obtained by crushing the rhizome in the morning for a week or till one is cured. For taste, a spoonful of honey may be added.
b. Woman who wants to terminate early conception of a child may drink about 50-60ml of the extracted liquid early in the morning in empty stomach till it ovulates.

How to Identify: The leaf structure of this *Alpinia galangal* species is similar with other *Zingiberaceas* species. It bears small reddish and white flower in bunches when matured the seeds bears like small greyish beads. Unlike others rhizome that grows underground this *Alpinia galangal* sprout upwards near the stem which is noticeable. They prefer to grow on sloppy areas, though cultivable. The dried residue of the crushed rhizome is preserved by the Chothes and neighbouring tribes to eat sometimes as chutney mixed with dry beef or boiled hides during off seasons as delicacy.

12. *Amaranthus retroflexus* Linn.

Family: Amaranthaceae

Common Name: Prickly/Spiny amaranth/Spiny pigweed

Local Names – Chothe: *Siak anbu (Aringtoipa)*; **Manipuri:** *Chengkrung (Tingkhangpanbi)*

Part Used: Leaf, Stem

Diseases/Treatments: Extra tissue growth/Haemorrhoid (Diabetic, Blood pressure).

Apply little lime (limestone) on the mother (first) extra tissue growth or haemorrhoid whether on the hand or leg. Then, when the tissue becomes little soft smeared repeatedly with the matured stem (stalk) along with the limestone paste against the extra tissue growth and by wasting it with water time to time. The process of applying lime paste, smearing it with the stalk of *Amaranthus spinosus* and washing with water is repeated till the tissues are broken and bleeds somehow. It is believed that if the mother tissue cell is completely uprooted the other tissues would miraculously fade away slowly within a week. The Manipuris called it *Sarik* because it grows in branches. It is not Keloid but looks alike.

How to Identify: This herb grows wild especially in dusty dry wasteland areas. It grows up to a high of 2-4 feet and the plant has small thorns. The other variety found in the region has no thorns. Some ate the tender leaves of both the species with dry fish as simple cooked curry. (There are different varieties used the thorny plant for this).

13. *Anisomeles indica* (L.) Kuntze.

Family: Lamiaceae

Common Name: Catmint

Local Names – Chothe: *Sipeng arapa*; **Manipuri:** *Thoiding angouba*

Part Used: Leaf.

Diseases/Treatments: Piles Boil some leaves and served half a glass twice a day for 3-5 days or till cured of the piles.

How to Identify: This herb looks like the linseed plant. It grows upto a high of 2-4 feet. The leaves are heart shape, soft and thick abit hairy with zig-zag on the outer sides. The flower blooms in cluster which is light purple along the seed pods. The smell of the leaves is grassy and taste slightly bitter. It is considered very effective for piles that ooze blood.

14. *Antidesma acidum* Retz

Family: Phyllanthaceae

Common Name: Amita.

Local Names – Chothe: *Tuitrit;* **Manipuri:** *Ching Yensin*

Part Used: Leaf

Diseases/Treatments: Indigestion and Enhance Immune System.

Person suffering from indigestion or wish to enhance immune system may drink half a glass of the boiled leaves as decoction anytime of the day (better after food), this enhanced the blood circulation and increase the immune system and indigestion.

How to Identify: This *Antidesma acidum* plant grows in shady areas among big trees of the tropical region. The leaves grow from both sides of the branch and it is little hard though smooth. It bears small white flowers. It is very difficult to identify among other similar plants. It normally grows up to a high of 3-7 feet. The fresh leave when chewed is abit sour. The Chothes and other hill tribes often gathers from forest area and is usually eaten as simple curry with dried fishes or meat. Sometimes, the leaves are boiled for awhile and when cold the water is squeezed out to reduce the sourness and served with meat chutney as delicacy. It is considered to be rich in minerals and a good agent of anti-oxidant. It is refreshing to have a glassful of the extracted liquid by adding little salt.

15. *Arisaema tortuosum* Schott.

Family: Araceae

Common Name: Whipcord cobra lily, Bagh jandhra

Local Names – Chothe: *Arui cheisu;* **Manipuri:** *Lin cheisu*

Part Used: Rhizome.

Diseases/Treatments: Virus infections/Skinny, Pricked by poisonous plants like boils.

Boil the fresh rhizome for a long time and let it cool. Then crushed the rhizome and bandage the paste around the infected wound area by the virus or bacteria. [Infection such as *Hin-naba, Lainingthougi tenna naba (lit. god's arrow), Lambi tingkhang (lit. thorn of footpath), Lai matu* (lit. gods hair) and *etc.* are treated with this herb].

How to Identify: The plant grows in shady or moist wetland areas with big split leaves and spotted stem or stalk which is dark green and yellowish (like the military camouflage dress colour). Some people like Garo tribe of Meghalaya, India fried and ate the stem/stalk of this plant. The species is very similar to *Arisaema tortuosum VanDusen* KT6FK1 (RM).

16. *Artemisia vulgaris* Linn.

Family: Asteraceae

Common Name: Mugwort, Indian wormwood, Sagebrush

Local Names – Chothe: *Leikha lei;* **Manipuri:** *Leibakngou akuppa*

Part Used: Leaf

Diseases/Treatments: Diabetes

Boil some leaves of *Artemisia vulgaris* and add little *sitamasi* and served half a glass of the decoction twice a day for 7 days or till cured.

How to Identify: The small flower is yellow in colour and the leaf is very bitter, hot and clumsy in taste. The leaf is dark green on the upper surface and whitish underneath. It appears like some sort of flower plants which are cultivated by some families in their courtyards.

17. *Averrhoa carambola* Linn.

Family: Oxalidaceae

Common Name: Star fruit

Local Names – Chothe: *Heinouchom;* **Manipuri:** *Heinoujom*

Part Used: Fruit.

Diseases/Treatment: (i) Mud stain on nails, (ii) Germicides.

a. The fruits or the extracted liquid is use to remove mud stain on finger nails by polishing with the fresh fruits especially by the local paddy trans-planters' during the season.

b .The fruit are crush and thrown into the polluted water and pond banks of fishery farm affected from severe flood to kill the germs and bacteria that produces very awful stinky smell.

How to Identify: The local variety plant grows upto a high of about 10-30 feet. The leaves are small, smooth, thin and heart shaped. The flowers are tiny and purple in colour. The fruits are long and have the regular star shape and very acidic in taste. Occasionally the extracted liquid are preserved and taken after food by mixing with water and sugar as digestive juice. The other variety of this star fruit is small and sweet while the big variety found in the hills is often very sour and acidic.

18. *Azadirachta indica* A. Juss.

Family: Acanthaceae

Common Name: Neem

Local Names – Chothe: *Theichak;* **Manipuri:** *Neem*

Part Used: Leaf and bark

Diseases/Treatments: (i) Cough, Skin diseases, (ii) Toothache.

a. Person suffering from cough and other skin diseases may boiled the fresh leaves in a pot of water and bath everyday till it gets cured.

b. The bark or twig of Neem is used in cleaning the teeth or for toothache in the past.

How to Identify: There are similar plant species. The leaves are small grows in opposite side from a long stalk of the branch. The flowers are small and white and purple in colour. The seeds are like big berries which is brown when ripe. The leave when crushed on palm has little awful smell. The genuine species of Neem plant is hardly available nowadays in the region.

Part B (B-C)

19. *Basella alba* Linn.

Family: Basellaceae

Common Name: Malabar spinach, Vine spinach, Buffalo spinach

Local Names – Chothe: *Noinotte;* **Manipuri:** *Urok sumban*

Part Used: Leaf.

Diseases/Treatments: Boil

Heat abit on the fire the fresh leaves and apply the soft leaves on the boil area with little opening at the tip of the boil. Repeat this process till the boil become ripe and about to burst out the pus. It is suggested that one should not mess with the boil until it is fully ripe for it becomes later cancerous and sceptic.

How to Identify: This creeper plant grows on fence around the house or at the roadsides or in the forest. The leaves are very thick, smooth and tender with heart shaped. The stem is brown in colour soft and watery. The seeds are black when ripe and often used as dye substance of blue colour.

20. *Bauhinia variegate* (L.) Benth

Family: Fabaceae

Common Name: Orchid tree, Mountain-ebony

Local Names – Chothe: *Chingthrao/Heina kundo;* **Manipuri:** *Chingthrao/Heina kundo*

Part Used: Leaf

Diseases/Treatments: (i) Anemia/Blood purification, (ii) Reduce weight.

a. Cooked some leaves as a simple curry along with fresh small shinny local fishes or prawn (Ngakha or Khajing) and eat twice a day (Morning and evening) for atleast 5 days to improve once health from weak and sickly conditions.
b. Perhaps, if the fresh leaves are cook as simple curry with dried small shinny local fishes (Ngakha) twice a day for 5 days or more, the person's weight gradually loses or decreases.

How to Identify: This is a shrub plant that grows upto a high of about 10-15 feet. The leaves are rounded and divided into two halves without pointed tips. The flowers are white-pinkish. The fruit bears like beans which is flat. However, if the leaves do not taste sour it is not the exact species that is referred for medication here rather it is an ordinary flower plant.

21. *Benincasa hispida* Thunb. Cogn.

Family: Cucurbitaceae

Common Name: Ash-gourd/Winter melon

Local Names – Chothe: *Maipoy/Mayngou;* **Manipuri:** *Torbot*

Part Used: Fruit

Diseases/Treatments: (i) Giddiness, Nausea, (ii) Chronic-ulcer, (iii) Snake bite.

a. Person suffering from frequent giddiness or nausea may apply the paste of this matured *Benincasa hispida* fruit above the forehead to sooth and calm down the blood pressure. The preparation method is that the outer green cover of the fruit is removed and the inner white fruit is sliced into small pieces (semi-paste form) with a spoon without touching ones hand for the required purpose. The paste is poured into a thin cloth and wrapped directly above the forehead.

b. When a person suffers from chronic-ulcer he may either eat fresh fruit in empty stomach each morning and evening or boiled and have it as simple cooked curry regularly till cured. Consider to be highly anti-oxidant.

c. Sabang (*Sirha* in Chothe) – horn is used to suck out the poison of the snake bite and after which the *Benincasa hispida* paste is applied on it.

How to Identify: This *Benincasa hispida* creeper plant is cultivated by most tribal communities of Northeast India for home consumption. The leaves are big and hairy. The flowers are also big and bright yellow in colour. The fruit which is like water-melon is covered with whitish ash powder. The local term it as white pumpkin. The seeds are like pumpkin seeds. The fruit is best served by Chothes as boiled or as culinary with duck and pork meat.

22. *Bischofia javanica* Blume.

Family: Phyllanthaceae

Common Name: Bishop wood.

Local Names – Chothe: *Uthum toh;* **Manipuri:** *Uthum panbi/Naraobi*

Part Used: Leaf (Tender).

Diseases/Treatments: High Blood pressure

The tender fresh leaves are either boil and eaten as simple boiled or the fresh leave is eaten along with local chutney along with food occasionally as food items to reduce the blood pressure.

How to Identify: The plant grows upto about 20-30 feet high. The leaves are thick, shiny and elongated, and tender when young. The flowers are tiny and yellowish blooms in clusters from the tip of the branches. It grows wild in the forest or cultivated by some family around their house as fencing.

23. *Brassaiopsis hainla* (Buch.-Ham.) Seem.

Family: Araliaceae

Common Name: Chuletro, Seto Chuletro Tree.

Local names – Chothe: *Chom* **Manipur**: *Chom* (Variety)

Part Used: Leaf.

Diseases/Treatment: Kidney failure, Urinary infection problems.

Boiled the fresh leaves in about 2-3 litres of water till it turns deep red colour and one may serve a glassful of the decoction twice a day for a week. In case of improvement the patient may continue to take the decoction solution till fully cured for a month. During the medication avoid any fresh red meat, oils, and spicy things.

One may also crush the leaves and take half a glass of the extracted liquid before food once a day for 3 days. If there is a sign of improvement continue for a week or till fully cured.

How to Identify: This is a shrub plant that grows upto a high of 10-20 feet. The stem and branches have thorns/spiky. The leaves have the shape of papaya but hard and rough and the leave stalk also has spikes. The flowers are small yellowish and the fruits bear in clusters of buds greenish which is eaten as chutney by the locals.

24. *Cajanus cajan* (Linn.) Millsp.

Family: Fabaceae

Common Name: Pigeon pea, Red gram.

Local Names – Chothe: *Meirongpi;* **Manipuri:** *Meirongbi*

Part Used: Leaf.

Diseases/Treatments: Dysentery

Crush some fresh leaves of *Cajanus cajan* and add little common salt and take about 20-30ml 2/3 times a day for 2-3 days.

How to Identify: This is shrub plant. The branches are brittle. The leaves are small, rough and elongated. The fruits bear pea pods in bunches. The seeds are the size of pea. The tender leaves and young fruit are clumsy in taste.

25. *Cannabis sativa* Linn.

Family: Cannabaceae,

Common Name: True hemp

Local Names – Chothe: *Ganja;* **Manipuri:** *Ganja*

Part Used: Leaf

Diseases/Treatments: (i) Diarrhoea, Dysentery, (ii) Lack of appetite, iii) Depression

a. Person suffering from diarrhea or dysentery may drink about 10-20ml of the extracted liquid obtained by crushing the leaves once a day after food till cured.

b. Person suffering from lack of appetite may fry few leaves mixed with some flour and taken some before food.

c. Person suffering from serious depression may smoke mildly the dried leaves in pipes or as cigarette. (It is suggested only if the person is completely free and willing to rest for few days from serious works).

How to Identify: This *cannabis sativa* is cultivated by some for personal used. The leaves grow in spit 3-4 petals forms. The leaves are long, tender and rough.

The plant bears small tiny white flowers and brown seeds. They prefer to grow in wasteland areas rich in humus. It grows upto a high of 4-7 feet. The leave is sticky when crushed on palm and give little awful smell. The use is not recommended for adolescences and young people.

26. *Capsicum annum* var. *fasiculatum* Baily.

Family: Solanaceae

Common Name: Bhut jolokia/Ghost chilly/Ghost pepper/Red Naga chilly/Naga Morich.

Local Names – Chothe: *Masi arokpa;* **Manipuri:** *U-Morok*

Parts used: Fruit

Diseases/Treatment: Ringworm

Cut fresh ripe king chilly into halves and apply on the ringworm area. Repeat this 3-4 times alternately till it is cured.

How to Identify: This king chilly or Naga chilly is widely grown in Northeast India. It is considered one of the hottest chillies in the world. This chilly belongs to shrub plant and grows up to a high of 3-4 ft. and lives for 3-5 years. This chilly have very thin cellulose outer covering and strong aromatic flavour. Locals preserved by drying in the sun and above the hearth the matured chilly for consumption during off-season. It grows well in high altitude.

27. *Carica papaya* Linn.

Family: Caricaceae

Common Name: Papaya

Local Names – Chothe: *Awathapi;* **Manipuri:** *Awathabi*

Part Used: Fruit

Diseases/Treatments: (i) Chronic ulcer, Gastritis and, (ii) Dog bites.

a. Person suffering from severe or mild chronic-ulcer may eat lots of papaya.
b. When a person is bitten by a dog immediately apply a slice of the green outer cover along with the pus on the bitten spot in order to neutralize or extract the poison and one may eat lots of unripe papaya too.

How to Identify: Different variety of papaya is widely grown in the region. The split leaves are huge with hollow stalk. When unripe the papaya is green and yellow when ripe and the seed are black. The plants grow upto a high of 7-15 feet. The local prefer to eat the unripe papaya along with chutney as delicacy besides as boiled.

28. *Cassia timoriensis* Wlld.

Family: Caesalpiniaceae

Common Name: Stout prickly climber

Local Names – Chothe: *Laiki;* **Manipuri:** *Laikee*

Part Used: Stem (Core part).

Diseases/Treatments: Jaundice, Kidney failure.

Boil for long time few small sliced of the dried stem (core part of the base stem) and when it turns blood red, cold it down and take 20-30ml of the decoction solution twice a day for 5 days after food. If there is a sign of improvement one may continue to take the prepared solution once every day and when required till fully cured.

For children suffering with jaundice below the age of 2-3 months old give 2-5 drops of the solution once a day only before feeding till cure.

How to Identify: This is a tree that grows upto a high of about 20-30 feet. The leaves are tiny smooth pinnate where the tips are slightly rounded grow on both sides of the stalk. The flower is yellow while the seeds pods are bigger than tamarind but short in length. The fruit is clumsy in taste. This plant may not be exactly the species of *timoriensis* though probably belongs to the genus Cassia. It is a rare species in the area.

29. *Celtis timorensis* Linn.

Family: Ulmaceae

Common Name: Southern nettle

Local Names – Chothe: *Heikreng;* **Manipuri:** *Heigreng*

Part Used: Leaf/Bark

Diseases/Treatments: Stone cases (Kidney)

Boil some fresh leaves and bark of *Celtis timorensis* and have the decoction a glassful 3 times a day till relief of pain in abdomen.

How to Identify: This black berry plant grows upto a high of about 15-30 feet. The leaves are elliptical, thin, and coarse when matured. The tiny berry fruits are pink-orange and black when ripe which is eatable. The birds are fond of this berry.

30. *Centella asiatica* Linnaeus. Urb.

Family: Apiaceae

Common Name: Asiatic/Indian Pennywort

Local Names – Chothe: *Anleiphon/Ariphon*; **Manipuri:** *Peruk*

Part Used: Whole plant.

Diseases/Treatments: (i) Hypertension, Gastritis, Chronic-ulcer, (ii) Constipation, Blood Purification, iii) Cough, (iv) Increase memory power, (v) Diarrhoea.

a. Persons suffering from serious hypertension, gastritis, chronic ulcer may take about 30-60ml of the extract liquid obtained from crushing the whole plant (leaf and stem). A spoonful of honey may be added or taken singularly as tonic before each meal regularly for a period of 1 week or more, depending upon the seriousness. If needed, the liquid may be diluted with some water and taken regularly.

b. Persons suffering from weak immunity, constipation or wished to enhance blood purification system may take a glass full of the boiled plant as concoction after or before meal or anytime for a week or more.

c. One may eat some fresh leaves time to time for 2 days or boil some fresh leaves and served a glassful of the decoction 3 times a day for 3 days.

d. Crush some fresh leaves and collect the extracted liquid in a clean container and add little honey in about 5-10ml and take the solution early in the morning in empty stomach regularly for month or occasionally time to time to enhance the memory power, especially for students. The dosages depend upon the age of the children or adult.

e. Crush some fresh leaves and add little honey and take about 10-30ml of the solution 3 times a day for 3-5 days.

How to Identify: The *Centella asiatica* creeper plant is grown wild on grassland areas either in the field or roadsides. The small leaves are slightly rounded in shaped with long stalks. Those grown on fertile wetland areas have long stalks while those grown in dry areas have short stalks with dark green leaves. It is slightly bitter and clumsy in taste. It is rich in iron and other minerals and served as a good anti-oxidant.

The Chothes and other tribals whether for health purpose or for its good taste often collects from the open fields or grassland areas grown wild during the season and served as either simple boiled or eaten the fresh as chutney.

31. *Cinnamomum tamala* (Buch-Ham) T.Nees

Family: Lauraceae

Common Name: Indian bay leaf/Tajbat

Local Names – Chothe: *Tastepat;* **Manipuri:** *Tastepat*

Part Used: Leaf

Diseases/Treatments: Urinary infectious problem

Boil some fresh tastepata leaves till it turns red and add little *Sitamasi* (crystal cane sugar) and serve about half a glass or glassful of the decoction solution once a day for 3 days or till cured. (It is suggested to take immediately when the irritation just began. One may also take the solution instead of water when demanded).

How to Identify: This plant grows upto a high of about 20-30 feet in the deep forest or planted in the farm. The leaves are shiny, smooth, long and hard. The leaves when crush produces a scent aromatic smell. Traditionally, the leaves are dried in the sun and preserved for future used in meat curry and in tea for its aroma.

32. *Cissus adnata* Roxb.

Family: Vitaceae

Local Names – Chothe: *Kokngoujen;* **Manipuri:** *Kokngouyen angouba*

Parts used: Leaf

Diseases/Treatment: Stone cases (Bile/Kidney/Urinary).

Boiled some fresh leaves of *Cissus adnata* along with some piece of Upupa epops (Hoopoes) bird's meat and served it as stew or simple boiled curry for a week to cure from any kidney/liver/bile's stone cases.

How to Identify: This climber plant grows well in cold shady areas. The leaves are heart shaped, tender and not so thick. The flowers are small and white in colour. The leaf taste little slightly sour.

33. *Citrus* × *aurantifolia* Linn/Burm

Family: Rutaceae

Common Name: Common/Sour lime

Local Names – Chothe: *Champra;* **Manipuri:** *Champra/Chambra*

Part Used: Fruit

Diseases/Treatments: Kidney (Stone) problem.

Person suffering from kidney stone problem may drink the mixture of fresh ripen extracted lemon juice with a local hen egg in empty stomach in the morning. The preparation method is that, the fowl's egg is dip into the glass half filled with the extracted lemon juice for about 12 hours. Then, the un-dissolved materials like the cellulose of the egg and bit of shell may be removed slowly and after which stirred well the mixture and drink it in empty stomach early in the morning. Later after 10-20 minutes, one may drink lots of water before anything is taken. This process may be repeated and taken till one gets relief of the pain/stone.

How to Identify: Preparation for this medicine required highly concentrated type of lemon which is locally grown in the hills by the tribals. Often the covering of this lemon grown in the hills has coarse thick covering unlike the hybrid which is smooth and thin and less acidic. Besides, the local fowl egg is preferred. One should note that after about 6 hours if the egg shell does not dissolve when rotating the un-dissolved upper egg shell it may not be used at all, signifying that the lemon is of low grade quality. Have the solution only if the egg shell fully dissolved after 6 hours. This process may be repeated every day or alternately depending upon the size of the kidney stone or urinary problems.

34. *Citrus maxima* (Burm.) Merr.

Family: Rutaceae

Common Name: Pomela/Grape fruit/Citrus grandis

Local Names – Chothe: *Nobab;* **Manipuri:** *Nobab*

Parts used: Fruit

Diseases/Treatment: Conjunctivitis

Apply 1-2 drops of the ripe fruit taken from the bigger extreme corner of the fruit capsule on the sore eyes or person easily infected with conjunctivitis.

How to Identify: This citric fruit is of orange family. This fruit is much larger than orange. It is found in two varieties - red and white in colour. The fruit is widely grown in tropical areas of Northeast India.

35. *Clerodendrum colebrookianum* Linn. Moon.

Family: Lamiaceae

Common Name: Bleeding heart

Local Names – Chothe: *Anphui (Soknoudon);* **Manipuri:** *Moirang Khanum*

Part Used: Leaf

Diseases/Treatments: Hypertension and Controls high Blood Pressure.

Persons suffering from hypertension, anxiety and high Blood Pressure may take a glass full of the fresh boiled leaves of *Clerodendrum colebrookianum* after each meal or anytime for 2-3 days or till the pressure reduces. However, prolong or excess intake of this decoction is considered bad as it lower the blood pressure again. Or the patient may consume some amount of the boiled leaves along with other food items occasionally.

How to Identify: The leave of this shrub plant is broad, smooth, tender and have the heart shaped. The flower of this eatable species is white and bloom in clusters, while the seeds when matured becomes red and turns black. It grows up to a high of about 6-10 feet. The leaves when crushed produced little grass smell. There are two types of species found in the region *i.e.* white flower which is eatable and the other with red flowers is not eatable at all. The Chothes and other hill tribes often ate the leaves as culinary item either as boiled or with pork curry as delicacy during the season of April- August. It is suggested that those suffering from low blood pressure may avoid eating it.

36. *Costus speciosus* (J. Konig.) Smith.

Family: Costaceae

Common Name: Crape ginger, Spiral ginger.

Local Names – Chothe: *Oakchak khombi*; **Manipuri:** *Oakchak khombi*

Part Used: Stem

Diseases/Treatments: Conjunctivitis, Sore eyes, Ear liquid.

Crushed the clean stem and add 2-3 drops into the sore eyes or ear infection side 2-3 times a day for 7 days or till it is cured.

How to Identify: The leaves are like turmeric but small in size. The flower is dark red and at the base of the centre it is blackish in colour. It grows wild in tropical regions sparely. It is also known as *Cheilocostus speciosus* (J. Konig.). There are different varieties some bears white flower and others red.

37. *Cretaeva religiosa* G. Forster.

Family: Capparidaceae

Common Name: Temple plant, Sacred garlic pear. (Syn: Crataeva nurvala Buck.-Ham.)

Local Names – Chothe: *Panthing lei;* **Manipuri:** *Loiyumba mapal/Loiyum lei*

Part Used: Leaf/Root

Diseases/Treatments: Joint pains

Boiled some amount of the fresh leaves and when cold, bandage the leaves around the joint pains like knee or hips, *etc*. Repeat this process after 8 hours or the next day till it is relief and cured.

How to Identify: The Chothe called this plant as fencing flower plant. It grows upto 20-30 feet high. The leaves are soft, smooth and shiny when young. The flower is white with 4 petals and the pistils in the centre are long and dark purple in colour.

38. *Curcuma caesia* Roxb.

Family: Zingiberaceae

Common Name: Black turmeric/Black zedoary

Local Names – Chothe: *Aisan ahan;* **Manipuri:** *Yaimu*

Part Used: Rhizome

Diseases/Treatments: (i) Smallpox, Tumour, (ii) Stomach ache, (iii) Dispeller of Magical Spell.

a. When a person suffers from smallpox or minor tumour the crushed rhizome is mixed with little mustard oil and is applied on the smallpox area or around the tumour. This procedure is repeated till the disease is cured. The patient (if feels) may even eat some amount of the fresh or dry rhizome about 10-20mg after food.

b. When there is sudden stomach ache chew a slight piece of the rhizome and take little water. One can take the little bit of the dried powder form also along with water.

c. Traditionally, some local medicine man/priest asked a person to drink the extracted crushed liquid of this rhizome mixed with little water and salt after offering ritual when considered suffering from the magical spell and charms. This tonic is taken for 3-4 days any time of the day. This Curcuma caesia rhizome (ginseng) is considered very rare to find.

How to Identify: Though belong to Zingiberaceae family the species is very rare in the region and is highly valued for its medicinal properties. The leave is similar with the turmeric plant, yet the rhizome is very small and blackish or greyish in colour. It has different aromatic smell and very pungent taste. The white-purple flower is very small and blooms for only a week or so. It also has religious significant. (Patient with low blood pressure should avoid taking this leaves).

39. *Curcuma longa* Linn.

Family: Zingiberaceae

Common Name: Turmeric

Local Names – Chothe: *Aisan;* **Manipuri:** *Yaingang*

Part Used: Rhizome

Diseases/Treatments: (i) Carminative Stimulant, (ii) Intestine Worms, (iii) Cuts and wounds, (iv) Complexion.

a. About 20-30ml of the extracted liquid obtained by crushing the raw rhizome of *Aisan* and by mixing with little water and salt may be taken by person suffering from indigestion, weak abdomen muscle bowel and carminative problems. This tonic should be taken before food in empty stomach atleast 3-4 times a day for a week. However, it is recommended to have once or twice in a month as health precaution since it helps in strengthening the bowel muscles and stimulates the digestive systems.

b. This turmeric tonic is also given to children believed to suffer from infested intestinal worms like tape worms, ascaris, *etc.* 2-3 times in a week.

c. When a person suffers from a major cut or wound on any body parts, the fresh turmeric is crush and the paste mixed with little mustard oil and lime (limestone) is applied on the cut area and bandaged loosely for 2 days. Repeat it again if there is improvement. Because of its special chemical property, it acts as a good agent of blood clotting and antiseptic. The wound healed faster with dry skin.

d If the fresh turmeric is slightly roasted and eaten as chutney regularly it is considered to enhance one's complexion and free from pimples.

How to Identify: The turmeric is cultivated by most tribal communities for home consumption in their farm or garden. Sometimes the fresh leaves are used in preparing a special cuisine by wrapping the local small fish mixed with chilly and salt, where the package is roasted under the hot ash under fire or steamed above the rice. The turmeric also has religious significant. The Chothe believed that the colour and odour of *Aisan* can dispel evil spirits. Therefore, they applied on their forehead while going in unknown forest areas at night time. They kept it below the pillows along with few garlic pods when feared from nightmares or to avoid from bad dreams.

40. *Curcuma zedoaria* (Christm.) Roscoe.

Family: Zingiberaceae

Common Name: White turmeric, Kentjur

Local Names – Chothe: *Aithing angoupa;* **Manipuri:** *Sing angouba*

Part Used: Rhizome.

Diseases/Treatments: Avoid nerves constrain, Stomach ache (To ward off evil spirit).

Chew a small piece of the rhizome early in the morning to relief from stomach ache and to ward off from evil eyes and evil spirits so that nothing happens to you like stomach ache or head ache on the journey.

How to Identify: The plant grows like wild turmeric plant with big leaves. But the smell of the rhizome has strong aromatic ginger smell, so it is locally known as white ginger. It is very bitter in taste instead of being pungent like turmeric or ginger. The flower is white/purple-yellow like turmeric flower that bloom at the base of the plant.

41. *Cuscuta reflexa* Roxb.

Family: Convolvulaceae

Common Name: Giant dodder

Local Names – Chothe: *Jamrui anemte/Awanrui;* **Manipuri:** *Kunjarata*

Part Used: Stem.

Diseases/Treatments: Diabetes (Urine case), (ii) Jaundice.

a. Crushed a bit and boil some amount of this yellow creeper plant in about 2-3 litres of water till it turns deep brown. When it gets cold take a glassful three times a day after food for a month or so. It is suggested that the decoction solution should be taken instead of normal water.

b. Boil some of the creeper along with big water bug (*Esing Nausek*), ghun seed and a Longpal seed and serve a glassful of the decoction solution 2 times a day for 7 days or more. The solution can also be used for bathing, washing hands and face occasionally.

How to Identify: This is a yellow creeper plant (*Kunjarata*) found grown around the fencing, riversides and trees. The small tiny white flowers have yellow pistil in the centre and bloom in clusters. The unique of this herb is that it grows hanging from other trees and has no roots. It gets its nutrient from air and other plants. It is suggested to seriously avoid alcohol (Liquor), tobacco, or any intoxicant substances during medication with this creeper.

42. *Cynodon dactylon* (L.) Pers.

Family: Poaceae

Common Name: Doob/Bermuda-grass

Local Names – Chothe: *Sanahuplang/Tingthou/Bangru-bange;* **Manipuri:** *Tingthou*

Part Used: Leaf/whole plant.

Diseases/Treatments: (i) Urinary Problem, (ii) Children stomach ache, (iii) Cancer, (iv) High Blood Pressure

a. If a person suffers from urinary problem, take about 2 tea spoonful of the crushed young tender leaves of *Cynodon dactylon* (*Tingthou)* mixed with little water and honey early in the morning before food till cured.

b. Some priests also suggested the extracted liquid of the crushed leaves and stems to children suffering from intestinal or stomach problems to give about 5-10ml twice a day for a week before meal.

c. Crushed some amount of the fresh leaves of *tingthou* and add about 2-5ml of the extracted ginger liquid and take about 10 ml twice a day for a month and if helpful one may continue on.

d. Boiled a handful leaves of the Cynodon dactylon and *Polygonum odoratum* (Phakphai) in 5 litres of water and collect only about 1 litre and served the decoction half a glass twice a day before meal till pressure reduced.

How to Identify: This is another common herb/grass found in the courtyard or grasslands areas of the state. It is creeper grass. The small elongated leaves grow in opposite directions from the stem at each knot. Some local priest used for other sicknesses too and has religious significances. This creeper herb grows wild on the courtyard or garden or lawn. The leaves are tender and grow from the knot of the stem one after another. The leaves are thin, small and long and bear a bud by some matured plants. It is believed to have cooling agent.

Part B (D – K)

43. *Dichrocephala integrifolia* (L.f) Kuntze

Family: Asteraceae

Common Name: Dicrocefala in Italian, Phakchee doi in Thailand.

Local Names – Chothe: *Aralu;* **Manipuri:** *Lalukok*

Part Used: Whole plant.

Diseases/Treatments: (i) Enhance blood (Post-natal woman), (ii) Shampoo.

a. Cooked the tender shoots/leaves as simple boiled curry along with fresh local fishes (*Ngara*) and serve occasionally for a week or two. This is highly recommended for the post-natal woman to enhance their blood considered to be rich in iron and other minerals.

b. Some fresh leaves are added in local made shampoo of rice starch and orange cover. It is considered that when this local shampoo is used regularly it keeps the head cold. The hair also becomes black, thick, smooth and shiny.

How to Identify: This herb grows wild on the courtyard, roadsides, and wasteland areas. The leaves are little thick and coarse. The white-yellowish bud like flowers blooms in cluster at the tip of the plant. The leaves is little clumsy in taste.

44. *Dipterocarpus tuberculatus* Roxb.

Family: Dipterocarpaceae

Common Name: Khmer khlong, Gurjan, Teak

Local Names – Chothe: *Khangra;* **Manipuri:** *Khangra*

Part Used: Leaf.

Diseases/Treatments: Menstrual problems.

The tender leaves are boiled and served as simple cooked curry with dried fishes for 3-4 day or occasionally till the menstrual problem is cured.

How to Identify: This plant grows wild in the forest and dry land areas. The leaves are big, triple and coarse or hairy. The flowers are bright to deep red and blooms in cluster at the tip of the long stalk. Traditionally, the leaves are used for wrapping food packages in olden days.

45. *Drymaria cordata* (L.) Willd.

Family: Caryophyllaceae

Common Name: Tropical chickweed, Heartleaf, Drymary.

Local Names – Chothe: *Tandan tadit;* **Manipuri:** *Tandan tadit*

Part Used: Whole plant.

Diseases/Treatments: Typhoid, Internal inflammation of stomach, Crack of lips.

Crush the whole plant and add little sitamasi (crystal cane sugar) and take half a glass twice a day for 2-3 days before food especially, early in the morning.

How to Identify: This creeper plant is found grown on wasteland area in cold temperate zone. The leaves are tiny round and the flowers are tiny white in colour. It is abit sour in taste. The locals sometime used for cleaning utensils in earlier days.

46. *Dysoxylum densiflorum* (Blume) Mig.

Family: Meliaceae

Common Name: Indian white cedar, Bili devdari, Bombay white cedar, Velley agil, Porapa

Local Names – Chothe: *Mishribu;* **Manipuri:**

Part Used: Leaf. **Diseases/Treatments:** Cancer, Tumour

Crush the fresh leaves and apply about 5 ml of the extracted liquid solution on the affected cancer or tumour area once a day till cure.

How to Identify: The tree grows in the deep forest upto a high of 20-40 feet. The plant bears tiny white flower and the leave is elongated and dark green in colour while the back side of the leaf is whitish. The leaf taste very bitter. The Rongmei (Kabui) of Manipur called it as *Jaiphung*. The species is synonymous with *Dysoxylum gaudichaudianum* and has similar leaf structure with *Toona ciliate* M. Roem.

47. *Eclipta alba* (Linn.) Hassk.

Family: Asteraceae

Common Name: Trailing eclipta

Local Names – Chothe: *Uchi sumban;* **Manipuri:** *Uchi sumban*

Part Used: Leaf

Diseases/Treatments: Inflammation of stomach, Crack on the lips and tongue.

Crush the fresh leaves and apply the extracted liquid time to time on the crack lips and tongue.

How to Identify: This *Eclipta alba* herb grows wild on the moist wasteland areas. The plant grows upto a high of about 1-2 feet. The flowers are tiny and white in colour with thick bud.

48. *Elatostema lineolatum* Wight.

Family: Urticaceae

Common Name: Lined Elatostema, Ediew, Hentekhan

Local Names – Chothe: *Tuidaitoh;* **Manipuri:**

Part Used: Leaf

Diseases/Treatments: Gastritis, Ulcer, Liver

Boil the tender leaves for about 5 minutes and have the boiled leaves in empty stomach and one may also have the decoction twice a day for 5-7 days to reduce from gastritis, ulcer and liver problems as it cools down the stomach from acidity. One may also cook along with rice and meat and have as stew regularly or occasionally to reduce acidity.

How to Identify: This herb is like a sub-shrub plant that grows in cold moist shady areas especially at the bank of the rocky streams. It grows upto a high of about 1-3 feet. The leaves are oblong, soft and tender. The white flower blooms in bunches from a single style. Even the stems are tender, greenery and watery. If over cook it is slippery. Most locals like Mao tribe often cooked along with rice and meat as stew or *palou* or *Kachari* type. The species is synonymous with *Elatostema tumidulum* Koidz. The leaf of uneatable variety is rough and hairy, even the stems/ stalks are hairy and hard.

49. *Elsholtzia graffithii* Hook.f.

Family: Lamiaceae

Common Name: Jangli tulsi/Ban tulsi/Mauhri, Lengmesel.

Local Names – Chothe: *Lengtu;* **Manipuri:** *Lomba*

Part Used: Leaf and flowers

Diseases/Treatments: Asthma, Tonsillitis, Nervous spasms.

Person suffering from asthma and nervous spasms may take about 2 tea spoonful of the extracted liquid obtain by crushing the leaves of *Elsholtzia graffithii* and adding some water twice a day before each meal, till it gets cured.

How to Identify: The leaves are small and oval shaped. The greyish flower is sponge and seed are tiny black. It grows about 1-2 feet in high. This *Elsholtzia griffithii* plant is cultivated in kitchen garden by most tribal communities. The dried leaves and dried spongy flower buds are preserved and added as flavour in chutney, fish, beef and yam curries as delicacies. The species is synonymous with *Elsholtzia blanda* Benth and *Dysophylla communis.*

50. *Elsholtzia ciliata* Thunb Hyland.

Family: Lamiaceae

Common Name: Vietnamese Balm, Jangli tulsi, Chinese mint bush.

Local Names – Chothe: *Areng;* **Manipuri:** *Teckta*

Part Used: Leaf and flower.

Diseases/Treatments: (i) Worm (babies), (ii) Bad old gas of stomach, (iii) Cough and cold fever.

a. Smear some fresh leaves of *Elsholtzia ciliata* with your palm and add little common salt and pure oil and apply the paste solution as balm on the baby's body (except the stomach area) to children infested with intestinal worms.

b. Chew some fresh leaves and drink some water 2 times a day for 3 days or more.

c. Smear some fresh leaves on your palm and add little common salt and pure mustard/olive oil and massage the whole body with the ointment as balm especially the feet, chest, hands, back and around the nostril area. But it is strongly suggest avoiding the massage around the stomach and belly areas.

How to Identify: The leaves are elongated and coarse. The flower is similar to *Elsholtzia blanda* species, but less sponge and seed are tiny. It grows about 3-5 feet in high. This plant is cultivated in kitchen garden by most tribal communities for home consumption. The dried leaves and flowers are preserved and added as flavour in chutney, fish, beef and yam curries as delicacies. The leaf has strong aromatic smell and slightly pungent.

51. *Entada phaseoloides* (L.) Merr.

Family: Fabaceae

Common Name: St. Thomas bean

Local Names – Chothe: *Kaang;* **Manipuri:** *Kangkhil* (*Kang*)

Part Used: Seed.

Diseases/Treatments: Scare mark, Bruise.

Keep smearing slightly the dried rounded seed 2-3 times a day to eradicate the bruise or scare mark by spelling or saying subside and get rid of the scare mark 3-5 times.

How to Identify: The seed of a climber plant. Traditionally this seed is used by the Meitei's for playing a game called 'Kang-sanaba' introduced by the Royal king Loitongba of Manipur. The climber plant is synonym of *Entada gigas* (L.) flawc. & Rendle.

52. *Eryngium foetidum* Linn.

Family: Apiaceae

Common Name: Saw tooth

Local Names – Chothe: *Awa-neem (Somey)*; **Manipuri:** *Awa-phadigom*

Part Used: whole plant

Diseases/Treatments: (i) Liver and Hepatic problems, (ii) Paralysis, (iii) Cuts and wound, and (iv) Sprain, (v) High blood pressure.

a. Person suffering from liver and hepatic problems may take about 10-15ml of the extracted liquid of the leaves of *Eryngium foetidum* obtained by crushing and adding a tea spoonful of honey regularly before each meal for a period of 1- 2 weeks.
b. The extracted liquid of the crush fresh leaves of saw tooth plant (mix with little honey) is taken about 20-30ml 3 times a day for 2 weeks. If there is sign of improvement continue taking the solution till cured.
c. Crush some fresh leaves and apply the paste on the cuts or wounds of the fowl or chicken's leg or neck.
d. Crush some fresh leaves and apply the paste around the sprain part and bandage it for 2 days. Repeat this process till fully healed.
e. Crush some of the fresh leaves and take about 10-20ml of the extracted liquid 2 times a day or one may boil and take a glassful of the decoction 2 times a day to reduce hypertension and high blood pressure.

How to Identify: The herb *Eryngium foetidum* is grown wild under the shade of trees or cold moisture grassland area of the tropical region. Nowadays, because of the demand some people have started to cultivate. The leaves are like saw tooth. The matured plant bears tiny bud like flowers and seeds from the long hard stalk coming out from the middle of the plant. The Chothes often served best the fresh leaves especially, in beef and pork curry to give better aromatic taste besides, adding in certain meat chutney.

53. *Erythrina variegate* (alba) Linn.

Family: Fabaceae

Common Name: Indian coral tree, Lenten tree, Tiger's claw.

Local Names – Chothe: *Hungthu angoupa;* **Manipuri:** *Kurao angouba*

Part Used: Bark

Diseases/Treatments: (i) Dispeller of evil spirit, (ii) Possess by evil spirit.

a. Cut a piece of the bark and polished it well into small pieces and is often wear around the neck of a child or on the arm band or waist of an adult to ward off evil spirits by most locals including the Meiteis (Manipuris).
b. Cut one of a big branch or stem and make a drum shell when dried by curving out a hollow, and cover and tide the two openings with the vulture's skin/hide. Perhaps, when the drum is beaten the evil spirit comes out and dances to the tune. (This is a belief that may perhaps involve magico-religious ritual).

How to Identify: The soft wood plant grows upto a high of about 15-30 feet, which is spiky all around. The leaves are thin, medium size and triplet in heart shaped. The flowers are whitish or crimson, long and pointed, blooms in cluster at the tip of the branch. The other common variety is red flower colour. This white flower plant has religious significant, especially for children and in child naming ceremony by the Chothe. The red variety is known as *Erythrina indica* Linn. The flower blooms in late winter or early spring when all the leaves shed.

54. *Eupatorium cammanoi* Linn.

Family: Asteraceae

Common Name: Hempagrimony

Local Names – Chothe: *Aripung/Renglei;* **Manipuri:** *Langthrei*

Part Used: Leaf

Diseases/Treatments: (i) Acidity, Gastritis and, (ii) Burning Sensation (Apigastric).

a. Person suffering from frequent acidity or gastritis/gas formation in the stomach may take about 10-20ml of the extracted liquid of the leaves obtained by crushing in empty stomach before each meals or atleast three times in a day. For children one may dilute the liquid with little water in a glass mixed with little salt and have it.

b. One may eat few fresh leaves and drink a glass of water if a person suddenly suffers from burning sensation of stomach from excess of eating chilly or sudden development burning sensation. In certain cases, one may have it in tonic form as described above for immediate relief from the pain.

How to Identify: This *Eupatorium cammanoi* herb is considered a very sensitive plant. It is usually planted in the courtyard or kitchen garden. The leaves are elongated and bear tiny flowers. It grows upto a high of about 2 feet. It is also used in certain offertory rituals. The leaves are bitter and clumsy in taste. It is an anti-inflammation and anti-oxidant plant. It is very sensitive to pure and polluted hands while plucking the leaves. The plant lived usually a maximum of three years and therefore, needs to be replanted again.

55. *Ficus hispida* Linn. F.

Family: Moraceae

Common Name: Hairy fig/Devil fig

Local Names – Chothe: *Theichang;* **Manipuri:** *Heirik*

Part Used: Leaf

Diseases/Treatments: Ringworm and Scabies.

Person suffering from ringworm may rub the leaf against the ringworm infected area or on the scabies part still it bleeds little. Later one may wash it with clean water and apply some pure (mustard) oil around it. This may be repeated 3-4 times till cured.

How to Identify: The local hairy fig tree is about 15-20 feet high and grows in the forest or roadsides. The leaves are broad, big and very coarse like sandy paper. The fruits are also little rough, rounded in structure. Brownish when unripe and turns red when ripe. It bears in cluster from a string of root like stalk usually at the base of the stem. In the past, usually people use to crush the unripe figs prepares chutney with mixed with (*Handak*) a kind of preserved pounded fermented dry fishes. The fruit served as a good antioxidant.

56. *Ficus racemosa* Linn.

Family: Moraceae,

Common Name: Cluster fig tree, Indian fig tree, Goolar

Local Names – Chothe: *Athi theichang;* **Manipuri:** *Asi Heibung*

Part Used: Leaf, Fruit.

Diseases/Treatments: Diabetes

Chew some fresh tender leaves and young fruits of *Ficus racemosa* 2 times a day for 1-2 months regularly.

How to Identify: This is big wild fig tree that grows upto a high of about 20-40 feet. The usually grows in deep forest or roadsides. The leaves are big, thick and rough. The fruit is like a fig but consider uneatable when ripe.

57. *Ficus tsjahela* Burm. Hook.f.

Family: Moraceae

Common Name: Pilkhan, Karal fig, South Indian fig.

Local Names – Chothe: *Abung;* **Manipuri:** *Khongnang tarung*

Parts used: Leaf

Diseases/Treatment: Diabetes

Boiled some leaves of *Ficus tsjahela* in about 3 litres of water and have a glassful of the decoction twice a day regularly and occasionally till reduced of the blood sugar level.

How to Identify: The plant externally looks similar with the Peepal plant but the leaf and plant of this variety is little small in size. The tender leaf which is brown and pinkish in colour tasted little sour and clumsy. Traditionally local plucks and ate the tender leaves as simple boiled curry with dried fish and meat or as local chutney as delicacies during the season of Feb-April. Unlike peepal and Banyan tree the aerial roots are absent. The leaves are shiny. The fruits are like tiny fig fruits deep brown to dark in colour when ripe.

58. *Gomphrena celosioides* Mart.

Family: Amaranthaceae

Common Name: Bachelor's button, Prostrate Globe-Amaranth.

Local Names – Chothe: *Noonte arapa;* **Manipuri:** *Chini kang*

Part Used: Whole plant.

Diseases/Treatments: Stone cases

Boiled the whole plant and served the decoction solution a glassful twice a day for 5-7 days or till relief of the pain.

How to Identify: This herb grows wild in the courtyard, dry wasteland areas or paddy fields. The elongated leave, stem and white flowers are hairy. The leaves are thick and soft. The flower grows from a bud of a long stalk.

59. *Goniothalamus sesquipedalis* (Wall.) Hook. F. & Thomson.

Family: Annonaceae

Common Name: Goniothalamus

Local Names – Chothe: *Aham;* **Manipuri:** *Laikham*

Part Used: Leaf.

Diseases/Treatments: Fever, Stomach ache (esp. children to dispel evil spirits).

A piece of the dried leaf is burnt, and the ash collected in a glass and little water is added. Drink little amount of the mix solution and the rest is applied on the whole-body parts. This is usually done twice or so only when a child suffers from sudden fever or stomach upset with a belief of contact of evil spirits. If the fever does not subside, then, it is considered a normal fever and other medication are given.

How to Identify: The plant grows in the deep forest in cold moist area or near the stream side. It grows upto a high of about 2-3 feet. The flower is orange in colour. The leaves are big, thick, smooth and well pattern. The dried leaves when burn produces a strong scent smell which the local belief that the smell can dispel and ward off the evil spirits. When children suddenly suffered from fever, stomach ache and other related symptoms they burn the dried leaves inside the house, even the burnt ash mixed with little water is given to the patient.

60. *Gynura cusimbua* D. Don/S. Moore.

Family: Asteraceae

Common Name: Hill gynura.

Local Names – Chothe: *Muhun lou;* **Manipuri:** *Terapaibi*

Part Used: Leaf

Diseases/Treatments: (i) Cuts and Wounds, (ii) Local shampoo (*Chinghe*)

a. When a person suffers from cut and wound apply immediately the crushed leave paste to stop from bleeding and help blood clotting. The plant is also acts as antiseptic.
b. The fresh leaves of *Gynura cusimbua* and pumpkin tender leaves (*Amai bu/ Maireng mana*) are added in the rice starch and left it to soak for 3-7 days after which the liquid is used to wash the hair to keep the head cool and hair smooth and silky.

How to Identify: This *Gynura cusimbua* is grown wild on fields and roadsides. The leaves are elongated and smelly. The red-white flower bud has cotton like fibres. Some Chothes are fond of eating fresh leave along with chutney. The Rongmei (Kabui) called this plant as *Jadonang*.

61. *Hedychium coronarium* (J.Koening)

Family: Zingiberaceae

Common Name: White garland –lily, White ginger lily or Yellow ginger

Local Names – Chothe: *Aidou;* **Manipuri:** *Yaingang ariba*

Part Used: Rhizome.

Diseases/Treatments: (i) Irregular menstrual problem, (ii) Gastritis or joints and muscle pain due to internal bad gases (*Dadhu nungsit tharoupa*)

a. Crushed the rhizome of *Hedychium coronarium* and add little common salt and take the solution of the extracted liquid about 5-10ml twice a day for 5 days before food in empty stomach early in the morning.

b. Boil the compound items of some *Aidou* rhizome, ghun seeds, Longpal seed, Jaiphor seed, garlic, local onion buds, a crab and white buffalo fats and take about 20-30ml twice a day for 3-5 days. One may also apply the solution on the joint pain areas 3-4 times a day for 3-5 days.

How to Identify: This herb belongs to Zingeberaceae family of turmeric. Locally, it is known as big or old turmeric as it gives more a stronger aromatic smell. The flower is white blooms in cluster. The leaves are abit hairy, thick, and slender, and it closely resembles the species *Zingiber cassumunar* Roxb. Traditionally, the Chothe used a piece of the matured rhizome as beads to wear around the child's neck as necklace to protect from any kind illnesses and to ward off against any evil spirits from harm and sicknesses.

62. *Hedychium flavum* Roxb.

Family: Zingiberaceae

Common Name: Ginger lily

Local Names – Chothe: *Soklei/Iphung;* **Manipuri:** *Loklei*

Part Used: Stem (young shoot)

Diseases/Treatments: Irregular menstrual problem

Boil the young shoots in some amount of water and take half a glass twice a day for a week to cure from irregular menstrual problems.

How to Identify: The plant is very similar to *puleimanbi*. However, the new young plant or the rhizome which is bright pink-reddish in colour is eaten as chutney occasionally during the season for its aromatic flavour by the locals of Manipur. Even the young flower buds are eaten as chutney for its aromatic flavour.

63. *Hedyotis auricularia* Linn.

Family: Rubiaceae

Common Name: Eared-Leaf Diamond Flower, Muttia lata, An Dien Tai, Gatekola

Local Names – Chothe: *Langpan Koukha;* **Manipuri:** *Langban Koukha*

Part Used: Stem, Leaf

Diseases/Treatments: White discharges of woman.

Boiled the leaves and stem and take a glassful of the cold decoction solution 3 times a day for 5 days in empty stomach.

How to Identify: This herb is short and has ovate leaf and grows wild in tropical wetland areas. It bear small white flower. The plant is hairy. The species is synonymous with *Oldenlandia auricularia* (L.) K. Schum. and *Exallage auricularia*

64. *Hibiscus acetosella* Welw. ex. Hiern.

Family: Malvaceae

Common Name: Cranberry hibiscus, African rosemallow, red leaved hibiscus.

Local Names – Chothe: *Anthui ashangpa;* **Manipuri:** *Sougari anganba*

Part Used: Leaf

Diseases/Treatments: Boil, Boil on the back (*Khoi mahum* in Meitei).

Boil the fresh leaves till it turn into paste form, then apply the paste over the infected boil area on the back side of the body. Crushed the leaves and make into paste and apply on the boil areas.

How to Identify: This is shrub plant often cultivated for home consumption. This red hemp plant has red flower so also the leaves and stems which red-brown or pale red in colour. The leaves and the flower petals are sour in taste. It is often eaten as simple cooked curry with dried fishes in summer season.

65. *Hibiscus cannabinus* Linn.

Family: Malvaceae

Common Name: Green sorrel/Kenaf/Deccan Hemp/Roselle

Local Names – Chothe: *Anthui angoupa;* **Manipuri:** *Sougree (ashinba)*

Part Used: Leaf and seed

Diseases/Treatments: Jaundice, Enhance Immune Systems and Energises the body.

Person suffering from jaundice may boil the fresh leaves of *Hibiscus cannabinus* or along with the soft outer covering of the seeds in about 1 litre of water and take half a glass as decoction twice a day anytime. Even the cooked leaves may be eaten.

How to Identify: There are two common types found in Manipur *i.e.* white and red species. The leaves are spread like fingers with a long stalk from the knotted stem or branches. The flower is big, white and dark purple in the centre, while the red/ maroon variety is also big. The green fresh leave that is abit sour is eaten as delicacy in summer as simple cooked curry with dried fishes. In the past, the Chothes and other tribal groups dried and preserved the leaves and soft outer covering of the seed for future to serve as simple cooked curry or eaten along with chutney with dried fishes and meat. Occasionally eating fresh leaves enhanced body stimulation or immune systems. The species is synonymous with *Hibiscus sabdariffa* L.

66. *Hibiscus rosa-sinensis* Linn.

Family: Malvaceae

Common Name: China rose, Hawaiin hibiscus.

Local Names – Manipuri: *Juba kushum*

Part Used: Root.

Diseases/Treatments: Irregular menstrual cycle of woman.

Boil some of the roots of *Hibiscus rosa-sinensis* and take a glassful 2 times a day till cure of the irregular menstrual cycle or problems.

How to Identify: This shrub grows upto a high of about 10-15 feet. The leaves are elongated with zig-zag on the outer part. The flower is deep red with multiple bushy petals. The species belongs to china-rose family, but this flower plant has no long pistil.

67. *Holskioldia sanguine* Retz.

Family: Lamiaceae

Common Name: Chinese hat plant, Cup and Saucer.

Local Names – Chothe: *Leiningthung asanpa;* **Manipuri:** *Kharam leisok anganba*

Part Used: Leaf/flower.

Diseases/Treatments: Boils, Body swelling due to bacterial infection, Scabies.

Boiled the fresh leaves of Holskioldia sanguine for a while and bandage the leaves around the swollen infected area 2 times a day till cured. It is suggested to avoid eating any fermented things like local dry fishes (Ngari, handak, hawajar) during medication. Crushed flowers are placed above the head to cure headache and giddiness by women in the past.

How to Identify: This is a bushy shrub plant. The leaves are elongated and smooth. The stalks are soft woody and have some spikes. The beautiful rounded pink or red colour flower blooms in bunches at each knots in the month of September-December. It is grown in the hilly area under shady plants. This flower has historical connection with the Pakhangpa the first historical king of Manipur, in which he brought this beautiful flower from one of his territorial expansion and gave it to his beloved wife Sunurembi while Pakhangpa (Thangwai Khongding) was the king of

Moirang. After the event, a decree was passed that no one should wear this flower except the queen. The flower was grown around the settlement of Kharam tribe when settled around Thangjing peak. The Meitei term it as 'Kharam leisok anganba or Toukhamlei'. The Maring called it Waa-aryam. There are two types - red and white variety. Both varieties are now endangered plants. Some local believes that who cannot conceive a child if offered this flower to their deity would conceive. It is a taboo to step the flower.

68. *Houttuynia cordata* Thunb.

Family: Saururaceae

Common Name: Chameleon plant, Lizard tail, Heartleaf.

Local Names – Chothe: *Anleithaang;* **Manipuri:** *Toningkok*

Part Used: Leaf and Stem (Root)

Diseases/Treatments: Gonorrhoea

Persons suffering from gonorrhoea may take about 20-30ml of the extracted liquid obtained by crushing the leaves and stems and adding little common salt before each meal twice a day for a week or till one gets cured.

How to Identify: This *Oroxylum indicum* is cultivated for home consumption as chutney. This herb grows well in a wet or shady area in the tropical areas. The leaves are heart shaped and the roots are tiny fibres while the stems are white in colour with several knots like sugar cane plant. The leaves and stem (root) when crushed has strong aromatic smell. The Chothes and other Northeast tribes eat this herb along with chutney made of local dry fish and other ingredients.

69. *Hydrocotyle sibthorpioides* Lam.

Family: Araliaceae

Common Name: Lawn marsh pennywort, Khulkhuri

Local Names – Chothe: *Lai peruk;* **Manipuri:** *Lai peruk*

Part Used: Leaf.

Diseases/Treatments: Stone cases

Crush the fresh leaves and add little water and take about half a glass twice a day for 5-7 days. (One is suggested to drink plenty of water during this medication).

How to Identify: This creeper herb grows wild or is cultivated by some people. The leaves rounded and are evergreen. It grows well in moist cool areas. The leaves taste little bitter and clumsy.

70. *Impatients balsamina* Linn.

Family: Balsaminaceae

Common Name: Garden balsam, Rose balsam.

Local Names – Chothe: *Khutshang;* **Manipuri:** *Khutjang*

Part Used: Leaf.

Diseases/Treatments: Cuts and wounds, Bruise, Wart.

Crush the fresh leaves and immediately bandage it the paste on the cuts and wounds especially, the bruise finger nail when hit by a hammer or hard substance.

How to Identify: The plant grows upto a high of about 2-3 feet. The leaves are elongated and grows knot after knot. The flowers bloom in cluster which is pinkish in colour. The seed pods are oval shaped and when matured it burst itself or when touch and curved automatically.

71. *Ipomoea batatas* (Linn.) Lam.

Family: Convolvulaceae

Common Name: Sweet potato

Local Names – Chothe: *Alu Mangkara;* **Manipuri:** *Mangara (Alu mangara)*

Part Used: Leaf

Diseases/Treatments: Giddiness/Migraines, Headache (esp. woman).

Crushed some fresh leaves of *Ipomoea batatas* and add little pure oil like Same, olive oil, and apply the paste above the forehead. This will shoot the head by keeping it cold thereby reducing the giddiness or dizziness or headache or migraine often occurred in elderly woman.

How to Identify: Any variety white or red species of the sweet potato leaves can be used. This is a creeper plant the rhizome or fruit grows underground along the roots. The leaves are heart shaped dark green in colour.

72. *Ipomoea carnea* Jace.

Family: Convolvulaceae

Common Name: Bush morning glory, Morning glory tree, Sedge plant

Local Names – Chothe: *Lamkhop lei;* **Manipuri:** *Sambal-pambi/Lamkhop lei*

Part Used: Leaf

Diseases/Treatments: Menstrual problem of white discharge.

Boil some of the fresh leaves along with along with the whole plant of *Kangphang ikaithabi* (*Mimosa pudica* Linn.) by adding little amount of sitamsi (crystal cane sugar) in about 2-3 litres of water and drink a glassful of the decoction solution twice a day in the morning and evening before food for a week or more. For those who are sensitive to the solution they got cured. (One should exclude the flower in the preparation).

How to Identify: The plant is a shrub that grows wild in the waste wetland areas to a high of about 5-15 feet. Some planted it as fencing plant. The plant's stem is brittle and hollow. The leaves have heart shape, soft and fatty. The flowers are medium size like hollow gramophone and are purple in colour.

73. *Isodon ternifolius* D.Don.

Family: Lamiaceae

Common Name: Kudo/Isodon nagarkot

Local Names – Chothe: *Koichu;* **Manipuri:** *Khoiju*

Part Used: Leaf.

Diseases/Treatments: Fever, Ward of evil spirit.

The dried leaves are burnt as incense to ward or wade of evil spirit when a person suddenly falls sick. The ashes are also applied on the forehead and body parts.

How to Identify: The shrub plant grows upto a high of 3-7 feet. The leaves are thick, rough and have white hairs underneath. \The flowers are tiny whitish and blooms in cluster at the tip of the plant. The stems are brittle and the dried leaves when burnt have scent smell. Traditionally, the Chothe used to burn the dried leaves and stem for religious purposes to wade away the evil spirits.

74. *Jatropha curcas* Linn.

Family: Euphorbiaceae

Common Name: Physic nut, Barbados nut, bubble bush.

Local Names – Chothe: *Awa keke;* **Manipuri:** *Awa keke*

Part Used: Leaf/Stem Gum

Diseases/Treatments: Cut, deep scratch and wounds

Collect some amount of the leaf of *Jatropha curcas* or stem gum in its leaf and apply immediately on the cut, deep scratch or wound and leaf it to coagulate the blood. It acts as first aids medicine. When heal it leaves no wounded/scare marks.

How to Identify: This shrub grows upto a high of about 10-15 feet. The plant is brittle and hollow inside and produces white gum on the stem or leaf knot or fruit knots. The leaves arė heart shaped with pointed tips at sideways. The flowers are yellow in colour and the fruit or seed bud are rounded with four sharp lines. This shrub plant is usually planted as fencing of the courtyard of a house in the villages for its survival longevity.

75. *Justicia adhatoda* Linn.

Family: Acanthaceae

Common Name: Adulsa/Malabar nut

Local Names – Chothe: *Anha/Chikpa,* **Manipuri:** *Nongmankha anganba*

Part Used: Leaf/flower

Diseases/Treatments: (i) Cough and Cold, Rheumatism, Feverish, Muscular/Joint pains, and (ii) Termination of Conception (Abortion).

a. For those who are suffering from slight cough and cold, and muscular/ joint pains some leaves are added in hot boiling water and is taken down after few minutes. Served half a glass the decoction (or add little common salt) twice a day after each meal or atleast three times a day till cured.

b. About 50-60ml of the extracted liquid from the fresh leaves is drunk early in the morning in empty stomach by a girl or woman who wants to terminate early conception of a child or for abortion. If required they may also have in the evening when the stomach is empty, for a period of 1 week or more. (This is considered a secret treatment).

How to Identify: The leaves are long and smooth. The flower is yellow-red blooms in bunches. It grows up to a high of 7-10 feet or more. The Chothes and other communities of Manipur usually planted around the courtyard as fencing and plucks when required. The tender leaves and matured flowers are also dried and preserved and eaten when needed as simple curry with dry fishes. Some locals prefer to eat the tender leaves (along with flower) fried with pork fats as delicacy. The leaves are added just few minutes before the curry is taken down otherwise the curry become over bitter to eat. This bitter-sweet acts as body stimulant too.

76. *Kaempferia rotunda* Linn.

Family: Zingiberaceae

Common Name: Indian crocus.

Local Names – Chothe: *Leipak lei;* **Manipuri:** *Leibak lei*

Part Used: Leaf.

Diseases/Treatments: Hot head, high Blood pressure.

Boil the fresh leaves along with the local shampoo (*Chinghe)* and wash the head occasionally or alternately to relief from hot head and high blood pressure.

How to Identify: This herb plant grows wild or is cultivated around houses. Unlike the turmeric family the flower is white-violet-purple in colour that blooms from the ground before the leaves appeared. Therefore, the local term it as earth or mud flower. The leaves look like turmeric but the back side is mixed with purple or violet in colour. The flower blooms only during April-May for only 7 days.

Part C (L–O)

77. *Laggera alata* (D.Don).

Family: Asteraceae

Common Name: Oliver

Local Names – Chothe: *Phephe;* **Manipuri:** *Haochak*

Part Used: Leaf.

Diseases/Treatments: Cough (esp. for children)

Collect few good leaves and heat near the fire and when it becomes tender, smear it with your palm and apply the liquid on the chest, neck, around the nostril and finally leaving the moist leaves above the head. Repeat this process 3-4 times in a day for 2-3 days.

How to Identify: The herb grows upto a high of about 2-4 feet in open wasteland areas. The leaves are thick, tender, abit hairy and with saw tooth structure on the outer part. The leaf is sticky when smeared and smells abit awful. The flowers are spiky, white-pinkish-purple in colour. In all four sides of the stem there is deform leaf like structure.

78. *Lantana camara* Linn.

Family: Verbenaceae

Common Name: Lantana weed, Big/wild sage

Local Names – Chothe: *Aring toh/Thirak lei;* **Manipuri:** *Nongban lei*

Part Used: Leaf

Diseases/Treatments: i) Blood Clotting, ii) Cuts and Wounds, iii) Freckle.

a. When a person suffers from minor cuts and wounds crushed some fresh leaves and immediately apply the paste on the cuts area. This will stop from bleeding as it helps in blood clotting and heal fast. The leave of this shrub plant is also believed to act as anti-fungal, microbial and insecticidal.
b. Crushed some fresh leaves of *Lantana camara* and rub with the paste on the freckles or red pigmentations area of the face and let it dry. Apply twice a day for a week. The red pigmentation will disappeared slowly.

How to Identify: This shrub plant grows wild upto to a high of 10-15 feet. There are two varieties found in the region one with white-pink flower and the other with red-yellow flower. The leaves are slightly elongated, coarse and rough. The four sides of the stems and branches have parallel thorns. Some locals said the tender leaves are eatable as boil or cooked as simple curry with dry meat or fish, but seldom taken. Most locals planted this shrub plant for fencing their compound. The dogs when having stomach upset usually eats this leaf to vomit.

79. *Leucas aspera* (Willd.) Linn.

Family: *Lamiaceae*

Common Name: Thumbai

Local Names – Chothe: *Wanchei ann;* **Manipuri:** *Mayang yenbum*

Parts used: Whole Plant

Diseases/Treatment: Blood enhancement (Post-natal woman), Sinus, Constipation.

Boil some fresh tender leaves as simple cooked curry along with fresh fishes and have regularly for 5- 7 days or occasionally to enhance blood especially for post-natal woman. It also suggested it helps cured sinusitis and constipation persons.

How to Identify: *The Leucas aspera* is grown wild in dry wasteland areas of the region. Occasionally the Manipur people ate the fresh leaves as simple cooked curry with dry fishes. The plant grows upto a high of about 1-2 feet. The leaves are elongated and hairy. The flowers are white and blooms in bunch at knots.

80. *Lindernia ciliata* (Colsm.) Pennell.

Family: Linderniaceae

Common Name: Hairy Slitwort

Local Names – Chothe: *Kihom maan.;* **Manipuri:** *Kihom maan*

Part Used: Whole plant

Diseases/Treatments: Stone cases (Kidney/Gall bladder)

Some amount of this creeper plant is boil in about 3 litres of water and taken a glassful 3 times a day for 5-7 days or till cured

How to Identify: This creeper plant grows wild on shady areas. The leaves are small, elongated with zig-zag like pineapple leaves at the outer sides or tooth. The flowers are tiny white in colour. This herb is synonymous with *Bonnaya branchiate* L. species.

81. *Litsea cubeba* (Lour) Pers.

Family: Lauraceae

Common Name: Mountain pepper, Litsea

Local Names – Chothe: *Sing-ngathai;* **Manipuri:** *Usingsha mapan*

Part Used: Fruit/flower

Diseases/Treatments: Asthma, Cool, Rhuematism.

Some amount of the fresh/dry fruit is crushed and eaten to eliminate asthma, cool and rheumatism once or twice a day or occasionally to dispel the unwanted oxidant elements in the body due to the pepper like components. About 2 ml of the extracted oil of the fruits is taken once or twice a day for a week or occasionally when required by those persons suffering from asthma, cool and rheumatism.

How to Identify: This tree grows upto a high of about 15-30 feet. It grows well in cool moist under thick shady trees in the forest. The leaves are smooth and elongated. The flowers are small white-yellowish in colour and bear fruit like berry. The young flower and fruit when young are eaten along with chutney by the locals since it is hot and gives aromatic taste. The fruit/seed are like when dried is like black pepper. If the seeds are eaten too much as chutney or added in meat curry it makes one feel sleepy or drowsy.

82. *Litsea glutinosa* sebifera Thumb.

Family: Lauraceae,

Common Name: Brown Bollywood, Soft Bollygum.

Local Names – Chothe: *Thaang hidak;* **Manipuri:** *Thaang hidak panbi*

Part Used: Bark, Root

Diseases/Treatments: Fracture and sprain

Crush the bark of *Litsea glutinosa sebifera* properly and bandage the gummy paste by adding little water around the fracture or sprain part for 2-3 days. Repeat the process by removing the old paste and applying afresh again after 2 days.

How to Identify: This is a tree that grows upto a high of about 20-30 feet or more. The leaves are thin and elongated. The barks are thick, when tasted or crushed well it is gummy and clumsy.

83. *Ludwigia adscendens* (L.) H.Hara

Family: Onagraceae

Common Name: Water primrose (white)

Local Names – Chothe: *Tui kundo;* **Manipuri:** *Esing kundo*

Part Used: Leaf.

Diseases/Treatments: Diabetes

Cooked the fresh leaves along with fresh local dog fish (*Ngakara*) as simple curry and serve it twice a day (Morning and evening) for atleast 3-5 days and may continue occasionally if helpful or till cured.

How to Identify: This *Ludwigia adscendens* creeper plant grows in the water bodies or near the watershed areas. The small leaves are thick and semi-rounded to elongated and soft. The flowers are tiny white in colour blooms at the knot of the reddish stem. The other variety which looks very similar with elongated leaf with yellow flower is not the species mentioned here.

84. *Lygodium japonicum* (Thunb.) Sw.

Family: Schizaeaceae

Common Name: Japanese climbing fern

Local Names – Chothe: *Lai changkhrang;* **Manipuri:** *Lai changkhrang*

Part Used: Whole plant.

Diseases/Treatments: Jaundice

Collect the ashes of the burn dried cree Practices of Ethnomedicines Among the Chothe of Manipur per plants and filter the ash water in a funnel and add little sitamai (crystal cane sugar) and take the cold decoction solution half a glass twice a day for 7 days or more after food till cured.

How to Identify: This is a creeper fern plant grown in the wetland areas on the fencing or wall of a house. The leaves look dry and are split in three array forms from a long stalk. It is also thin and appears pale yellow-green in colour. The creeper stem is bright yellowish, tender and watery if grown in moist area and little hard if found on dry land.

85. *Lysimachia candida* Lindl.

Family: Primulaceae

Common Name: Manipur loosestrife

Local Names – Chothe: *Kengoi;* **Manipuri:** *Kengoi*

Part Used: Whole plant.

Diseases/Treatments: Anaemia.

Boiled the whole plant of tender stems along with leaves and served as decoction half a glass twice a day for a week. One may also cooked as simple curry along with dried fishes and served occasionally.

How to Identify: This creeper plant is rich in iron contents and grows in the sandy paddy fields or around the wet-moist areas or sandy plane areas. The leaves are small oblong and bear tiny white flowers in cluster. The leaves are little sour in taste.

86. *Melastoma malabathricum* Linn.

Family: Melastomataceae

Common Name: Matakui (Palau)

Local Names – Chothe: *Joubung;* **Manipuri:** *Yachubi*

Part Used: Leaf.

Diseases/Treatments: (i) Jaundice, (ii) Diabetes

a. Boiled the fresh leaves and take half a glass of the decoction twice a day for 7 days after food only. It will be milky and sticky after boiling it.
b. Boil some fresh leaves and take a glassful of the decoction 3 times a day for a week or continue till the diabetes millipedes (frequent urine excretion) is reduce. It is suggested to have the decoction solution instead of water whenever one feels thirsty. This is for those diabetes persons suffering from frequent urine release.

How to Identify: The plant grows wild on roadsides or hilly areas upto a high of about 3-5 feet. The flower is white-pinkish-purple in colour. The tender leaves are cooked along with chicken as delicacy by some of the Chothes and hill tribes.

87. *Melothria perpusilla* (Blume.) Cogn.

Family: Cucurbitaceae

Common Name: Pabasari, Wild cucumber

Local Names – Chothe: *Lam sakma.* **Manipuri:** *Lamthabi*

Parts used: Whole plant/Tuber

Diseases/Treatment: Jaundice, Liver, Mal-nutrition.

Boil some of the whole plant either fresh or dried in about 2 litres of water and take a glassful of the decoction 3 times a day for 5-7 days to cure from jaundice. One may also take the decoction occasionally. Even the fruit are used in the preparation of the medicine. One can use the tuber when not available the leaf.

How to Identify: This *Melothria maderaspatana* creeper plant grows in the deep forest areas. The leaves and stem are hairy and spiky. The leaf is heart shaped. The flower is bright yellow. The fruits are slightly rounded and spiky like tiny cucumber.

88. *Menthe arvensis* Linn.

Family: Lamiaceae

Common Name: Spearmint/Pudina

Local Names – Chothe: *Mayang rilpum/Nungsi hidak*; **Manipuri:** *Nungsi hidak*

Part Used: Leaf

Diseases/Treatments: Stomach ache, Gastric, Antipyretic.

Person suffering from stomach ache, gastric and antipyretic cases may take about 10-20ml liquid of the crushed fresh leaves (or may add little honey) twice a day till cured.

How to Identify: The dotted leaves are oval in shaped. It grows up to a high of 2 feet. The local prefer to eat fresh leaves along with chutney. It is also eaten along with some certain sour fruits to reduce sourness or itchy teeth.

89. *Meyna laxiflora* Robyns.

Family: Rubiaceae

Common Name: Muyna/Helu/Alu

Local Names – Chothe: *Theipi;* **Manipuri:** *Heibi*

Part Used: Leaf and dry fruit

Diseases/Treatments: (i) Blood Purifier, (ii) Constipation.

a. To enhance blood purification of a person the fresh leaves may be eaten as chutney time to time, considering rich in iron contain.
b. If one is having constipation problem eat some of the ripe fruits.

How to Identify: The *Meyna laxiflora* plant is grown wild or is planted in the state. The leaves are elongated, smooth and shiny. The stem and the branches have pointed thorns, in which, sometimes the thorn is even used like needle to prick out foreign objects from the body parts. The fruit is like apple with big seeds inside. It is green when unripe and brown when ripe. It grows upto a high of 15- 25 feet. If one eats too much of ripe fruit it gives giddiness. Occasionally, the Chothes and Manipuris are fond of eating the leaves as chutney mixed with dried pea powder as delicacy. The Chothes used the leaves in certain village ritualistic offerings. The species is synonymous with *Vangueria spinosa* Linn.

90. *Millettia pachycarpa* Bentham.

Family: Fabaceae

Common Name: Fish poison.

Local Names – Chothe: *Ngamu jai;* **Manipuri:** *Ngamu jai*

Part Used: Stem, Root.

Diseases/Treatments: (i) Dog flea, (ii) Fish poison.

a. Cut a piece of the mature stem or root and split into two halves and wear it around the dog's neck. The dog's flea will get rid itself of the smell.

b. Crush some roots and put it in the river cleavages or stream, after half an hour or an hour the fishes will appear on the surface giddy struggling to survive.

How to Identify: This climber shrub grows in the deep forest. The leaves are long and coarse, and a bit thick when matured. The stems and roots are hard with fibres. Traditionally, the locals when goes for fishing in the stream or shallow river used this creeper plant and similar other plants as fish poisons to catch fishes. The flowers are whitish in colour.

91. *Mimosa pudica* Linn.

Family: Fabaceae

Common Name: Sensitive plant/(Touch-me-not)

Local Names – Chothe: *Ajakpi thingna;* **Manipuri:** *Kangphang ikaithabi*

Part Used: Leaf, root

Diseases/Treatments: Post-delivery Uterine pain, White discharge.

Woman suffering from post-delivery uterine pain may take about 10-20ml of the liquid obtained by crushing the leaves and roots as tonic (or mixed with honey) before meal twice a day till one gets relief. Sometimes the whole plant is boiled and served as decoction a glassful before food twice a day for a week or more till cured.

How to Identify: This *touch-me-not* plant grows wild especially in sandy and dry areas. The stem has thorns and the leaves are sensitive to any touch. It has ball shaped purple flowers. The plant is used with other plants to stop white discharged of woman.

92. *Mussaenda roxburghii* Hook. F.

Family: Rubiaceae

Common Name: East Himalayan Mussaenda, Soklati.

Local Names – Chothe: *Sunubarei;* **Manipuri:** *Hanurei*

Part Used: Leaf, Root

Diseases/Treatments: Boil, Cancer, Tumour.

Boil some fresh leaves and bandage around the boil or infected tumour or cancerous area for 3-5 days by dressing every day.

How to Identify: This shrub grows upto a high of about 5-10 feet. The leaves are well pattern, soft hairy and oblong. It bears some white leaves like white flowers at the tip of the branch and blooms small flower like bud in orange colour. The white flower species is commonly found in the region and the other violent-orange colour flowers species is hardly found. This white leaf flower species is considered valuable. The plant is also used in many other ailments by certain tribal communities of the state.

93. *Musa sapientum* Linn.

Family: Musaceae

Common Name: Banana stem/Pseudostem

Local Names – Chothe: *Changlong;* **Manipuri:** *Laphu and Laphu tharow*

Part Used: Stem, Flower

Diseases/Treatments: (i) Chronic ulcer, (ii) Lactation mother, (iii) Constipation, Diarrhea, (iv) White discharge of woman.

a. Person who suffers from prolonged chronic ulcer may drink about half a glass (60ml) of the extracted liquid of the edible young fresh banana stem (Pseudo stem) in empty stomach once a day for a week or more.
b. Young mother who lack breast milk may cooked along with some dried fishes and eat regularly for a week or frequently time to time to improve the breast milk for the baby. Therefore, this young banana stem especially grown from the hills are most preferred by the locals.
c. Some recommended that eating the simple cooked banana stem with dry fishes or as chutney helps reduced constipation if taken occasionally.
d. Cut the matured stem of the banana that is flowering already and extract the liquid of the middle stem stalk and take about about half a glass of it twice a day for 3 days. The taste is abit clumsy.

How to Identify: There are different varieties of banana. Among the eatable species the locals preferred the banana stem grown from the hill areas considering that the fibres are softer and less pungent, bitter and clumsy in taste. However, the people of Manipur including Chothes occasionally ate either as chutney after boiling the banana stem or adding it in pork curry as delicacy since it absorbs the fats.

94. *Nicotiana tabacum* Linn.

Family: Solanaceae

Common Name: Tobacco, Tamaku.

Local Names – Chothe: *Thakna;* **Manipuri:** *Hidak Mana*

Part Used: Leaf

Diseases/Treatments: Acts as insecticides, Itchy, Allergies, Insects bite, Leech bite.

Traditionally, when a person is bitten by an insect or leach, feels itchy, experience allergy one usually smears the tobacco leaf with water or spit and is apply on the infected area. The application of tobacco leaf on infected area acts as anti-allergy and repellant.

How to Identify: This short variety of *Nicotiana tabacum* is cultivated. The leaves are broad and coarse. It grows upto a high of 2-4 feet. It grows only for a short period. The leaves are dried in the sun and in the shades and preserved for future used. This dried leave is usually chewed along with betal nuts and leaves and sometimes a piece is kept between the lower lip and gum. The water is often spit out. It also acts as insecticides.

95. *Nyctanthes arbor-tristis* Linn.

Family: Oleaceae

Common Name: Night flowering Jasmine, Parijat.

Local Names – Chothe: *Singgharei;* **Manipuri:** *Singgarei*

Part Used: Leaf.

Diseases/Treatments: Malaria fever.

Crushed the fresh leaves and add little honey in the extracted liquid and take about 10-20ml twice a day for 2-3 days before food. (For children give less dose about 1-5ml). Avoid any fresh red meat during medication.

How to Identify: This is a shrub plant grow upto a high of 20-18 feet. The leaves are coarse, thin and oval shaped. The flower is white and reddish inside the base that bear rounded seed pods.

96. *Ocimum sanctum* Linn.

Family: Lamiaceae

Common Name: Holy basil/Krishna tulsi

Local Names – Chothe: *Tulsi;* **Manipuri:** *Tulsi*

Part Used: Leaf

Diseases/Treatments: Cough, Bronchitis, Tonsillitis, Asthma.

Person suffering from most air-borne diseases like cough, bronchitis, tonsillitis, asthma and congestion of lungs may take about 2 tea spoonful of the extracted leaves juice mixed with little honey 3 times a day till it is cured.

For children below the age of 3 years suffering from cough, add some fresh leaves in some amount of boiling water and give 2-3 tea spoon of the decoction 3 times a day till relief.

How to Identify: This Holy basil or Tulsi is widely found in the region. The leaves are small and the tiny flower is white-purple in colour. The average high is 2-3 feet. The fresh leaves when chewed is little hot. The Manipuris (Meitei) usually plant in the middle of their courtyard. The plant also has religious significant.

97. *Ocimum tenuiflorum* Linn.

Family: Lamiaceae

Common Name: Thai holy basil, Tulasi.

Local Names – Chothe: *Tulsi ahangpa;* **Manipuri:** *Tulsi amuba*

Part Used: Leaf.

Diseases/Treatments: Sinusitis

Crushed or smeared some fresh leaves with your palm and keep inhaling the smell time to time regularly 3-4 times a day for 7 days or till cured.

How to Identify: It is similar with the Ram Tulsi (Holi Basil) herb but the stalk and the leaves of this plant are blackish or deep purple in colour. The smell is lighter.

98. *Oenanthe javanica* Blume DC.

Family: Apiaceae

Common Name: Dropwort/water parsley/water celery

Local Names – Chothe: *Komprek;* **Manipuri:** *Kongprek*

Part Used: Leaf

Diseases/Treatments: Hypertension, high Blood Pressure, Indigestion

Person suffering from hypertension, high blood pressure and indigestion may take about 20-30ml of the extracted liquid of the crushed fresh leaves of *Oenanthe javanica* by adding little water (and honey) or half a glass after mixing with little water before food or after as tonic twice a day till cured. However, many preferred the decoction style of medication *i.e.* by boiling certain amount of the leaves and stem and serving a glassful twice a day before or after food and also eating the boiled leaves and stems along with other food items.

How to Identify: This *Oenanthe javanica* herb is grown near the watershed areas like near the river bank or pond side or lake side in the region. The leaves are small, long, tender and smooth. The flower is white bloom from a bud like form. It grows upto a high of 3 feet. There are different varieties but this short variety is most sought for. The stem is hollow inside with knots where leaves and branches grew. The people of Manipur are fond of eating rather raw the fresh leaves and stem as chutney (*Singju*) and cooked as simple curry with dry fish or dhal curry as delicacy. It is highly recommended for persons suffering from gastritis and hypertension to eat regularly as simple boil curry with dal and dry fishes.

99. *Oroxylum indicum* Linn.

Family: Bignoniaceae

Common Name: Indian trumpet flower

Local Names – Chothe: *Maklong;* **Manipuri:** *Samba/Shamba*

Part Used: Bark, leaf and fruit

Diseases/Treatments: (i) Piles, (ii) Muscles/Joint pains, (iii) high Blood pressure, (iv) Epilepsy, (v) Tumour and Cancer, (vi) Maggot.

a. Person suffering from piles may sit on the warm boiled water of the plant bark of *Oroxylum indicum* kept in a tub. This process is continued till one gets relief. The plant bark should be peeled upwards with one breath. Even the young tender fruit is slightly roasted and taken twice a day till the piles is cured.
b. Person who suffers from muscles/joint pains may take a bath with the boiled water of the bark or leaves. Or take a small towel and dipped into the boiled hot water of the leaves/bark, and balms the joint pain areas till relief.
c. Roast a piece of the young tender fruit in hot ash fire and eat twice a day for 3-5 days or till the high blood pressure is reduce. (Those suffering from low blood pressure may avoid taking it).
d. Those suffering from Epilepsy may either eat the tender leaves of *Oroxylum indicum* cooked as curry or the fruit slightly roasted occasionally till cured.
e. Crush a piece of the bark and apply the paste on the around the wounds or tumour or cancer areas for two days and repeat the process.
f. Crush the fresh leaves and apply the extracted liquid and bandage the paste on the infested area by the maggot, especially on animal's wounds. Repeat this process every day till cured.

How to Identify: The *Oroxylum indicum* tree is a soft plant. The multiple leaves on sideways from a stalk are heart shaped. The big white flower bears the fruit of long bean. Some people who like the taste eats the tender leaves as well as the young fruit along with chutney as delicacy.

100. *Oxalis corniculata/stricta* Linn.

Family: Oxalidaceae

Common Name: Creeping/Yellow woodsorrel.

Local Names – Chothe: *Yensin arikpi;* **Manipuri:** *Lam Yensin.*

Part Used: Leaf/Whole plant

Diseases/Treatments: (i) Indigestion, Gastric and (ii) Blood purification.

a. Person suffering from indigestion or gastric may take about 20-30ml of the extracted liquid of the whole plant of *Oxalis corniculata* before food by adding little water and honey or salt twice a day for a week or till it is cured.

b. Boiled the whole plant in about a litre of water and one may add little salt and take the decoction half a glass or full glass twice a day for a week or time to time to enhance blood purification. It is suggested that during medication one should avoid fatty or oilly things, chilly or spicy items.

How to Identify: This *Oxalis corniculata* creeper herb is grown in cluster wild in cold moist areas in the region. The stalk of a leaf has three split leaves and bears small bright yellow flowers. The fruits are tiny resembling the shaped of bitter gourd. The leave and fruit is abit sour. The other species is not recommended by the people.

Part D (P–R)

101. *Paederia foetida* Linn.

Family: Rubiaceae

Common Name: Skunkvine, Stinkvine, Chinese fever vine

Local Names – Chothe: *Yoisukjam;* **Manipuri:** *Oinum*

Part Used: Leaf.

Diseases/Treatments: Fracture (esp, of bird and animals)

Crush the fresh leaves and apply the paste especially animals and birds wounded area.

How to Identify: This is a creeper plant grown on the roadsides and fencing areas in the forest. The leaves are long and elongated, rough and very smelly like farted smell. The creeper is also very strong and has strong farted or awful smell. The tiny flowers are white and deep purple inside. Sometimes the local used the creeper as binding material for firewood or thatch. The species closely resembles *Sembukan Paederia foetida species.*

102. *Parkia timoriana* (DC.) Merr.

Family: Fabaceae

Common Name: Tree bean (Monkey rice)/Parkia Javanica

Local Names – Chothe: *Yongtak;* **Manipuri:** *Yongchak*

Part Used: Fruit

Diseases/Treatments: (i) Chronic ulcer, (ii) Dysentery, and Diarrhoea, (iii) Diabetes.

a. If one is suffering from prolong severe chronic-ulcer boiled some amount of the dried outer covering of the long tree bean fruits (*Parkia roxburghii)* in about a litre of water for half an hour or more till the water becomes deep red. When it is cold take half a glass of the decoction twice before food regularly for a week or more till it reduces the problem. The bitter and clumsy taste helps relief the pain and thickens the stomach and intestinal linings.
b. For person suffering from dysentery and diarrhea may also take half a glass of the decoction twice a day till relief of the problem. During medication one is suggested to avoid hot, spicy and oily food items.
c. Crush into fine powder form the dried young fruit (*Yongchak*) and add 2 tea spoonfuls in about a litre of boiling water and take the decoction solution half a glass twice a day after food for a week. It is suggested to drink the solution instead of water).

How to Identify: The plant *Parkia roxburghii* is widely planted in the state and is a delicacy for many. The leaves are tiny and grow in multi-small stalks in opposite sideways of the long stalk. The flower bud is bright yellow with white ring above it. The green fruit bears like long beans hanging from the tree. The plant grows upto a high of 20-30 feet. The fruits are scale off with sharp steel object like tongue-cleaner and the fruit/seeds are either cut in small pieces and eaten raw along with chutney or boiled and eaten as chutney in different forms. Often the green scales or outer covering of the fruits/seeds are thrown away as waste. The *Parkia timoriana* (DC.) Merr is synonymous with *Parkia roxburghii G. Don* and with *P. Javanca* Lam.

103. *Phaseolus lunatus* Linn.

Family: Fabaceae

Common Name: Butter bean, Lima bean

Local Names – Chothe: *Kalandri;* **Manipuri:** *Kalandri*

Part Used: Leaf.

Diseases/Treatments: Fever, Hot head.

Crush the fresh leaves of *Phaseolus lunatus* and apply the paste above the head and around the pituitary glands areas time to time till cured.

How to Identify: This is a climber plant. The leaves are heart shaped, hairy and pale. The flowers are small white-bluish or purple in colour. The seed pods are eaten as vegetables by the locals. Even the fresh tender leaves are cooked as simple curry with dried fishes and eaten during the season as delicacy.

104. *Phyllanthus urinaria* Linn.

Family: Phyllanthaceae

Common Name: Chamber bitter, Gripeweed, Stonebreaker, seed under the leaf, Gale of the wind.

Local Names – Chothe: *Lam theishura arapa;* **Manipuri:** *Chakpa heigru*

Part Used: Whole plant

Diseases/Treatments: Diabetes

Boil certain amount of the fresh whole plant and serve the decoction solution a glassful three times a day for a week. If there is a sign of improvement, prepare the decoction and take according to once requirement once a day or twice a day. The solution is a replacement of water. The diabetes patient should drink the solution instead of water whenever thirsty.

How to Identify: This herb is like a miniature or bonsai type of a gooseberry plant. It grows upto a high of about 1-2 feet. The structure is like miniature peepal tree, grows wild on roadsides or dry wasteland areas. The leaves and fruits are tiny. (Photos from website). The species is synonymous with *Phyllanthus amarus*, and *P. niruri*.

105. *Phyllanthus emblica* L.

Family: Phyllanthaceae

Common Name: Indian gooseberry, Amla

Local Names – Chothe: *Theichuraa/Shuru;* **Manipuri:** *Heikru/Heigru*

Part Used: Bark, leaf, fruit

Diseases/Treatments: (i) Cough and Cold, (ii) Sore Eye, (iii) Body stimulant (iv) Stone cases, Joint pains.

a. Person suffering from cough and cold, and those feeling weaknesses may crushed the fresh goose berries and take about 10-30ml 3-4 times in a day till relief for adult. For children below 5 years give 5-10ml only. For cough some prefer to roast abit the goose berry in the fire have as much as one can time to time till cured.

b. For sore eyes or conjunctivitis few drops of the extracted juice of the gooseberry may be poured into the sore eye 3 times a day or till one gets cured.

c. The extracted liquid obtained from the crushed fresh bark is mixed with little water and taken as body stimulant substance.

d. Boil the leave or the fruit and take certain amount like a glassful 3-4 times a day till relief of the abdomen and joint pains

How to Identify: The goose berry is widely grown in the deep forest while some planted around their courtyard. The leaves are tiny grows in small stalks from the branches. The flowers are also very small and bright yellow like pollen grains. The fruit goose berries bear from the twigs and branches which is yellowish when matured. Normally the plant grows to a high of 10-18 feet. Locally prepared goose berry pickles are commonly sold now.

106. *Plantago major* Linn.

Family: Plantaginaceae

Common Name: Broad plantain, white man's foot, Greater plantain.

Local Names – Chothe: *Antapot;* **Manipuri:** *Yampat*

Part Used: Leaf/stem

Diseases/Treatments: (i) Boils and wounds, (ii) Gastric, (iii) Jaundice.

a. Person suffering from boils may heat up the fresh leaves little on the fire till it gets decoloured. Then, roll abit and applied on the boil part with little opening on the tip. Repeat this process till the boil is matured and the pus comes out or healed. Crushed the fresh leaves and apply the paste on the wound. Repeat it time to time. Those suffering from urinary disorder may eat fresh tender leaves as simple boiled curry.
b. Boil some of the fresh leaves and take a glassful of the decoction 2 times a day for a month to cure from gastritis. (It is suggested to avoid spicy items like chilly, red meat and hot red tea).
c. Boil some fresh leaves and give bath with the decoction solution to the child suffering from jaundice regularly till cured.

How to Identify: The leaves are soft, tender and elongated. It grows wild on roadsides and cold areas of temperate tropical region. The flower and seeds bear in cluster from a long stalk. Occasionally, the tender leaves are also eaten as boiled or as simple cooked curry with dry fish.

107. *Polygonum odoratum* Lour.

Family: Polygonaceae

Common Name: Smartweed/Laksa plant/Vietnamese mint

Local Names – Chothe: *Phakphai;* **Manipuri:** *Phakpai*

Part Used: Leaf

Diseases/Treatments: (i) Hypertension, Tonsillitis and Throat problems, (ii) Indigestion, Hiccup, Old gases inside the stomach. (iii) High Blood pressure, Hypertension.

a. Person suffering from hypertension, tonsillitis and other throat problems may take about 10-20 ml of the fresh extracted liquid of the leaves of *Polygonum odoratum* mixed with little honey as tonic once a day for 3-5 days before food.

b. Chew few fresh leaves of *Polygonum adoratum* after food twice a day for 3-5 days to remove the bad indigestive gases from the stomach. This will help indigestion, remove bad gases and disappeared frequent hiccups.

c. Those suffering from hypertension and high blood pressure may boil some fresh leave and take 20-30ml 2 times a day for 3 days.

How to Identify: This Vietnamese mint herb is planted for home consumption in the kitchen garden for its strong aroma. The leaves are long, smooth and fibrous. The matured stem is hard. It grows upto a high of 1-2 feet. The tender leaves are often added in chutney for its aroma. Fresh leaves are added in fish chutney for its aroma and to enhance its taste. (A very similar type of this herb of knotweed/grass grows wild which is not eatable).

108. *Pouzolzia indica* (L.) G. Benn.

Family: Urticaceae

Common Name: Bo mam, Apoyapoyan, Graceful Pouzolzsbush

Local Names – Chothe: *Muithemte;* **Manipuri:**

Part Used: Tuber.

Diseases/Treatments: Boil, Cancer (Tumour on thigh, arm and muscle areas).

Crushed the rhizome or tuber of this *Pouzolzia indica* plant and apply the gummy paste around the boils. For cancer treatment it should be used only in the advance stage of the cancer tumour part) and bandage it for 2 days and repeat it again the process with fresh tuber till cured. The utensil used should be sterilised before used. (It is strongly suggested to avoid using any sharp metallic objects on the infection or while dressing it).

How to Identify: The leaves are hairy and oval shaped grows on opposite sides. The long thin stalk is also hairy and reddish in colour. The plant grows upto a high of 1-2 feet. The flower is small white-yellowish in colour. It grows in busy dry grassland areas. The Rongmei (Kabui) of Manipur called it as *Pheinap*. This very good herb is considered now as endangered plant because it is hardly available.

109. *Prunus persica* Linn.

Family: Rosaceae

Common Name: Peach **Local**

Local Names – Chothe: *Chumprei;* **Manipuri:** *Chumbrei*

Part Used: Leaf

Diseases/Treatments: Wounds and Worms, Anti-wormicide.

If a wound is infested with worms and maggots or became septic the fresh matured leaves of *Prunus persica* are crushed and the paste is applied on the wound (and the liquid may be squeezed into the wound before applying). This treatment is usually used on animals like pig, cow and buffalo when infected with maggots.

How to Identify: The peach plant is about 7-15feet high. The leaves are long, thin and tender. The fruit is hairy, green and yellowish when ripe with reddish colour around the stalk. It is planted in the courtyard as fruit plant.

110. *Psidium gajava* Linn.

Family: Myrtaceae

Common Name: Guava

Local Names – Chothe: *Bareitun;* **Manipuri:** *Pongdon*

Part Used: Leaf and fruit

Diseases/Treatments: Dysentery and Diarrhoea.

Traditionally, person suffering from dysentery or diarrhea may eat some fresh tender leaves of guava (with little salt) time to time till relief. Though little bitter and clumsy in taste it greatly helps in suspension.

Eat plenty of unripe guava fruit if one is suffering from dysentery and diarrhoea (with little common salt if needed be). However, excessive eating of unripe guava leads to constipation.

How to Identify: A wide variety of guava plant is widely grown in the state. Any variety is recommended for the purpose. The flower of guava is normally white with lots of pistils.

111. *Punica granalum* Linn.

Family: Onagraceae

Common Name: Pomegranate.

Local Names –Chothe: *Kapo thei/Theisan;* **Manipuri:** *Kaphoi/Kapohei.* **Parts used:** Leaf, fruit

Diseases/Treatment: Dysentery / Diarrhoea.

Chew the fresh tender leaves of the local pomegranate as much as possible to help suspend the diarrhoea or dysentery. One is suggested to keep chewing the fresh leaves whenever one feels like any time of the day.

How to Identify: The plant is cultivated in the courtyard by most families. The plant has some spikes on the stem and branches. The leaves are small, shiny, smooth and abit thick. The flower is bright orange colour and so also the fruit which is rounded in shaped with a big opening of the ovary at the tip. The seeds covering are bright red to dark red, although the seeds are light brown or white in colour. The maturity of the fruit is known by the drying up of the ovary or pistil.

112. *Rhus succedanea* Linn.

Family: Anacardiaceae

Common Name: Wax tree / Wild varnish dry galls

Local Names – Chothe: *Hokma;* **Manipuri:** *Heimang*

Part Used: Leaf / Fruit

Diseases/Treatments: (i) Dysentery, Indigestion, (ii) Constipation and (iii) Body stimulant.

a. Person suffering from dysentery and indigestion may eat some of the fresh tender leaves of *Rhus succedanea* (mixed with little salt) till it is suspended. Or about half/full glass of the boiled dry fruit of the decoction (mixed with little salt and sugar), when completely cold is taken 2/3 times a day till one gets relief.
b. On the other hand, those persons suffering from constipation may drink the grind seeds powder of *Rhus succedanea* mixed in hot water along with little chilly and salt, that served as enema.
c. When one feels lethargic or dull or lacks body stimulation add a tea spoon or two of the grind dry seeds powder in a litre of boiling water and served half/full glass of it. Or one may add a teaspoon or two while preparing red tea and also add sugar to one's taste and have either regularly or occasionally when needed to energise or enhance body stimulation especially after heavy meal.

How to Identify: This *Rhus succedanea* shrub grows upto a high of 10-15 feet. The elongated leaves are soft but not so smooth and grow on opposite sides of the stalk. The taste of the leaf is clumsy when chewed and the ripe fruit very sour or acidic. The fruit bears in cluster which is pink when young and turns snowy white in winter and becomes brown when fully ripe. Some people preferred to add some amount of the grind seed powder in fish curry to harden the fish and while in pork curry to reduce the fatty taste.

113. *Ricinus communis* Linn.

Family: Euphorbiaceae

Common Name: Castor bean, Castor oil plant

Local Names – Chothe: *Keke ashangpa;* **Manipuri:** *Keke anganba*

Part Used: Leaf

Diseases/Treatments: Giddiness (esp. woman).

Boil some fresh leaves along with the local shampoo (*Chinghe*) and wash the head and hair regularly for a week or occasionally with the solution to get relief from the giddiness or morning sickness type.

How to Identify: The shrub plant is grown wild on wasteland, roadsides or near the fencing of a house. The leaves are big, rough and splits like a star shaped grown from a long red stalk. The fruits are round like big marble with spikes all around. The stems are also red or maroon in colour. The other variety is green in colour.

114. *Rorippa Palustris* (L.) Besser.

Family: Brassicaceae

Common Name: Indian yellowcress, Indian Rorippa, Indian fieldcress.

Local Names – Chothe: *Aju antam;* **Manipuri:** *Uchi hangam*

Part Used: Leaf

Diseases/Treatments: Diabetes, Dysentery

The Fresh plant of *Rorippa Palustris* is cooked and eaten regularly against diabetes. The whole plant is boil and a glassful is served twice a day for two days those suffering from dysentery.

How to Identify: This foliage herb grows wild on moist or wetland areas to a high of about 1-2 feet. It is commonly found in the dusty fertile courtyard. What is commonly found in Manipur is the *Rorippa palustris* (L) in the wet wasteland areas rather than *Nasturtium officinale* R.Br. grown in water bodies - both belonged to same Brassicaceae family. This *Rorippa palustris* has bright yellow colour with 5-6 long tentacles bearing seeds.

Part E (S–Z)

115. *Sacharum officinarum* L.

Family: Poaceae

Common Name: Sugar cane.

Local Names – Chothe: *Molshu/Maanshu,* **Manipuri:** *Chu (anganba/angouba).*

Part Used: Stem.

Diseases/Treatments: Jaundice.

Person suffering from jaundice may drink 1 or 2 glassfuls of freshly extracted sugar cane juice 3-4 times daily, especially in the morning for 2-3 weeks till one gets cured from the jaundice.

How to Identify: Two common types of sugar-cane are widely prevalent in Manipur *i.e.;* the white species and red species. It grows upto a high of about 5-10 feet. Some believe the red sugar-cane juice is better than the white species for jaundice, while others considered white species is better. Nevertheless, both are good.

116. *Sapindus trifoliatus* Linn.

Family: Sapindaceae

Common Name: South India Soapnut, Soap berry.

Local Names – Chothe: *Lingsi,* **Manipuri:** *Kekru.*

Part Used: Fruit.

Diseases/Treatments: Fever (esp. children).

Soak half of the dried seed for a while in a half a mug of water and stir it with a spoon till it gives a thick foam then apply the foam above the head time to time till the fever subside and it reduces the body temperature.

How to Identify: This is a soft wood tree. The leaves are small, elongated and grow in multiples in opposite side ways from a long stalk. The seed when dried is dark brown or dark red in colour. In the past, the dried seeds are used as soap to wash clothes and utensils like brass and gold.

117. *Schefflera arboricola* (Hayata) Kanehira.

Family: Araliaceae

Common Name: Dwarf umbrella plant.

Local Names – Chothe: (*Tarang keikhomei* in Rongmei), **Manipuri:** *Utangbi.*

Part Used: Leaf.

Diseases/Treatments: Body/Muscle/Joint pains.

Boil the fresh leaves of *Utangbi* (nine leave variety), along with *Nungshangpanbi* and *Nongmangkha* and used the warm decoction solution as balm by giving a sponge body massage with it twice a day for 3 days or occasionally to relief from muscles and body ache.

How to Identify: This climber appears like a shrub plant when young. When it grows tall it clings to big trees for support. The leaves are thick, smooth and shiny like plastic coating. The flower is white-yellow. The plant is grown as houseplant for its decorativeness of the leave.

118. *Scutellaria discolor* Colebr.

Family: Lamiaceae

Common Name: Skullcap, Daamapaate/Nilo.

Local Names – Chothe: *Ann lakhat,* **Manipuri:** *Yenakha (Ching yenakha).*

Part Used: Whole plant.

Diseases/Treatments: (i) Irregular menstrual problems (removal of bad blood), (ii) Malaria, (iii) High blood pressure, (iv) Fever (esp. children).

a. Crush some fresh leaves and stems (whole plant) along with a piece of dry tobacco leaf and shock it with little pure alcohol (liquor) for two days, (plus few drops of tobacco liquid) after which, apply the concentrated liquid upwards from the feet and downwards from the head to shoulders, chest and back (but not around the stomach or belly area).
b. Boil the whole plant and add little honey and have about 5-10ml of the decoction solution 2 times a day for 3-5 days or till cured. (Avoid take more than the required amount). For children give about 1-2ml only.
c. Boil the fresh leaves along with black sticky rice and tobacco extracted liquid (*Thakpu tui*) and use the solution as balm from head to shoulder and feet to hip twice a day for 3 days or till reduced.
d. Crush the fresh leaves on your palm and add little saliva and apply the paste on the child's body, chest and back to reduce the fever.

How to Identify: This herb plant grows especially under the shady moist areas in riverside or thick forest. The leaves are oval shaped. The flowers are tiny white to bluish in colour that blooms in semi-cluster from a stalk from the centre of the plant. The leaves are extremely bitter in taste and it stays for long time in the mouth.

119. *Sida rhombifolia* Linn.

Family: Malvaceae

Common Name: Rhombus-leaved sida, Paddy's lucerne, Jelly leaf, Lalberela.

Local Names – Chothe: *Think ulin,* **Manipuri:** *Uhan.*

Part Used: Leaf.

Diseases/Treatments: Worms (esp. for children).

Crush the leaves and apply the paste on the baby's or child head. When the paste becomes dry due to body heat then, add few drops of water on the paste to just wet it. If not so effective repeat the process once again but not too frequently.

How to Identify: This herb grows wild on the fences, roadsides or dry wasteland areas like a small Peepal or Banyan tree structure about 1-3 feet high in fertile areas. The leaves are slightly small, thick and oval. The flowers are tiny bright yellowish. The stem and roots are hard to pluck therefore the local term it as old tree or senior wood.

120. *Solanum indicum* Linn.

Family: Solanaceae

Common Name: Yellow berried nightshade (small variety).

Local Names – Chothe: *Samtrok arikpa,* **Manipuri:** *Leipungkhang.*

Part Used: Fruit, Leaf.

Diseases/Treatments: (i) Piles, (ii) Measles and, (iii) Cough, (iv) Liver cases and Sore tongue, (v) Nail wounds.

a. The raw fruits of *Solanum indicum* is crushed and mixed with little honey and taken before each meal. Person with serious piles case that bleeds may take three times in a day, till it is cured.

b. Babies and children suffering from measles may take the grind fresh fruits mixed with little honey twice or thrice in a day till it completely expels out the virus and is cured. The locals believed that the measles should not be imbedded inside the body for it is deadly. (One should not give too much to young babies).

c. The young fruits are steamed and prepare as chutney or added in simple local dishes to prevent from cough and other ailments.

d. Crush some fresh fruits and add little honey and take about 5-10ml twice a day for a week before food to heal the infected liver and sore bright tongue.

f. Pluck seven leaves of this *Solamum indicum,* wash them clean and crush it into paste and apply on the wounded toe or finger nails that swells to rot for a day or two. It acts as antiseptic and heals the wound fast.

How to Identify: This *Solamum indicum* herb plant is normally planted in the garden for medicinal values. The plant and the leaves are spiky with thorns and grow up to a high of 4 feet. The flower is purple, and the pistil is yellow in colour. The fruit is green spotted when young and orange colour when ripe. The Chothes and other tribes sometimes collects the young green fruits and steamed above the rice or fresh green fruits are eaten as chutney, sometime the green fruits are added in simple plain local dish called *Outti.* Most people eat sometime to prevent from cough and other ailments.

121. *Solanum virginianum* Linn.

Family: Solanaceae

Common Name: Kantakari, Surattense nightshade, Yellow-fruit nightshade, Thai striped eggplant.

Local Names – Chothe: *Pheisouthai,* **Manipuri:** *Japan kamen.*

Part Used: Seed, Fruit (ripe).

Diseases/Treatments: (i) Tooth ache, (ii) Animals/Cow leach (*Lingaret*).

a. The seeds of the ripe fruits are extracted and kept it dry. Then burn the dried black seeds with the matches in a clean container and inhale the smoke through the mouth focusing towards the toothache. It is believed that the germs of the toothache fall down after a while. Repeat this burning the dried seeds and inhaling the smoke 3-4 times till the tooth ache is relief.
b. Crush some of the ripe fruits well and pour it into the animal's nostril like cow or buffalo infested with parasitic worm or leach. Within few minutes the infested leaches will exit itself from the animal's nostril.

How to Identify: The herb grows wild like weeds in wasteland areas. The big leaves and stem have sharp pointed yellow spikes or thorns which is extremely painful and scratchy. The fruits are green with stripes and turns bright yellow when ripe. The tiny seeds are black. The fruit are used as lubricants by children for their toy cars in earlier days.

122. *Tamarindus indicus* (L.)

Family: Caesalpiniaceae

Common Name: Tamarind.

Local Names – Chothe: *Mangke,* **Manipuri:** *Mangge*

Parts Used: Seed.

Diseases/Treatment: Measles.

Apply just 2/3 drops of the ripe fruit of tamarind fruit jam on the tongue and lips of the measles infected child. And later if the child demands for his craving items, provide and feed him to his satisfaction.

How to Identify: This *Tamarindus indicus* is grown in the family courtyard or gardens. The tree belongs to hardwood. The fruit bear like beans but very sour and turn brown in colour when ripe. The leaves are oblong and grow in parallel in multiples.

123. *Thalictrum foliosum* DC.

Family: Ranunculaceae

Common Name: Gold thread root, Pilijari.

Local Names – Chothe: *Changkrang aroupa.*

Parts Used: Leaf.

Diseases/Treatment: Sprain.

Crush some amount of the fresh leaves of *Thalictrum foliosum* into paste and apply it around the sprain or fracture part for 2 days and repeat the process till it is cured.

How to identify: This is creeper plant commonly grown in shady or moist loamy soil in the hills around the stony areas. It is perennial. The flower is small white in colour. The leaves are slightly oval with 2 zig-zags at the sides of the leaf.

124. *Toona ciliate* M. Roem.

Family: Meliaceae

Common Name: Red cedar.

Local Names – Chothe: *Atithing/Tairen,* **Manipuri:** *Tairen.*

Parts Used: Leaf.

Diseases/Treatment: (i) Typhoid, (ii) Cough, (iii) Malaria fever (iv) Measles and Small pox, (v) Allergy.

a. Crush some leaves of *Tairen mana* and add some honey and have a teaspoonful twice a day after food till the typhoid is reduced. (Pluck the leaf that has single leaf or three leaves at the tip of the stalk).

b. Boil some amount of the fresh leaves of *Toona ciliate* and served about 10-20ml of the decoction early morning in empty stomach, have 3 times a day for 3-5 days. One may also boil separately certain amount of the fresh leaves and take a bath with the decoction, if one does not have fever.

c. Boil some fresh leaves and take about 10-20ml of the decoction solution 3 times a day in empty stomach for 3-5 days in case of person suffering from malaria fever. (Do not take bath during fever).

d. Boil some fresh leaves of *Toona ciliate* and take about 10-20 ml in case of measles or small pox of the decoction solution 2 times a day for 3 days or more. (For children the dosage should be 1-5ml). After fully cured, one may take bath with the decoction solution boiled separately.

e. Collect some fresh leaves of *Toona ciliate* and dry fry it till it turns harden and blacken, then crush into powder form and add little pure mustard oil and apply the paste over the infected body parts twice a day for 2-3 days.

How to Identify: This is a tree that grows upto a high of about 20-40 feet. The leaves are long and grow multiple from a stalk in both sides. It belongs to a hardwood. Traditionally, the leaves are used to sprinkle the holy water to dispel evil spirit on certain religious ceremonies and death rituals.

125. *Trachyspermum roxburghianum* DC. Craib.

Family: Apiaceae

Common Name: Wild celery/Ajmod.

Local Names – Chothe: *Ching Pathikom,* **Manipuri:** *Ching padhigom.*

Part Used: Leaf/whole plant.

Diseases/Treatments: Enhance Blood Purification.

Person who wants to enhance blood purification may take about 10-20ml of the extracted liquid obtained by crushing the leaves mixed with little honey twice a day before each meal for a week or more till satisfied.

How to Identify: The herb is similar with the common celery, but this has stronger pungent aromatic smell than the common celery used in curry. It is seasonal herb and grows up to a high of 1 foot. The flowers are tiny white in colour. Those people especially the hill tribals who like the strong aromatic smell often planted at home for home consumption, though seldom used. The species is synonymous with *Carum roxburghianum.*

126. *Vitex negundo* Linn.

Family: Verbenaceae

Common Name: Chaste tree.

Local Names – Chothe: *Jenlik shibi,* **Manipuri:** *Urik shibi.*

Parts Used: Leaf.

Diseases/Treatment: i) Allergy, Scabies, ii) Chicken fleas.

a. Boil some fresh leaves and take a bath with the warm solution for 3-5 days to cure from allergy and scabies.
b. Stick some branches inside and around the poultry to dispel the chicken fleas. It acts as insecticides.

How to Identify: This shrub-creeper plant grows upto a high of about 5-10 feet. Being a soft plant, it bends like a creeper often. The leaves are small oval shaped smooth and shiny. The flowers are small and bluish in colour and bloom in cluster at the tip of the branches.

127. *Viola pilosa* Blume.

Family: Violaceae

Common Name: Viola, Banafshah

Local Names – Chothe: *Shantang leipak,* **Manipuri:** *Huikhong.*

Part Used: Leaf.

Diseases/Treatments: To enhance blood and stimulates body.

The fresh leaves are either boiled or eaten afresh along with local chutney occasionally to enhance blood and stimulate the body.

How to Identify: This herb grows in open moist grassland of hill areas or shady grasslands areas. The leaves are soft hairy, heart shape and the outer part is like saw tooth. It tastes little slipper and sour. The Chothes and other hill tribes occasionally gathered from the forest grasslands and ate it as simple cooked curry with dried fishes. The flowers are small white-pinkish in colour blooms in clusters from a long stalk. (This species may not be exactly *Pilosa,* since there are many varieties). The leaves are closely similar with *Viola primuifolia,* but the flower is different.

128. *Zanthoxylum acanthopodium* D.C.

Family: Rutaceae

Common Name: Pricky winged leaf/Darmar/Sheguan pepper.

Local Names – Chothe: *Singree,* **Manipuri:** *Mukthrubi.*

Part Used: Leaf and seed.

Diseases/Treatments: Cough, Asthma and Arthritis (Dry cough), Food Poison.

Person suffering from dry cough, arthritis and asthma may boiled the leaves and inhaled the hot steam and one may also take about 10-20ml of the decoction twice a day for a week or occasionally till relief. When leave is not available, one may boil some amount of the brown seed covering in half a liter of water and served about 10-20ml when warm to relief from dry cough and breathing congestion till relief. (It is suggested to avoid sweats and oily items during medication).

The compound mixture of *Zanthoxylum acanthopodium* leaves, *Tulsi, Longpal,* a slice of ginger and crystal cane sugar (Sitamasi) are boiled in about 5 litres of water and only 1 litre is collected, and about 30-50 ml is taken twice a day before meal for a week.

Crushed some fresh leaves and eat as much as possible immediately to relief from food poisoning of lethargy.

How to Identify: The leaves grow on opposite sides that are elongated and have small spikes underneath. The stems and branches are also spiky with thorns. The seeds when young are green like small berries and become pinkish to red when ripening and turns deep brown when dried. Actually, what is eaten is the leaves and seeds covering which has strong spicy aromatic smell, while the seeds are black in colour and often removed. Because of the aromatic smell the Chothes and other hill tribes often add the fresh leaves and dried seed shells in dog and beef curry giving a good flavour of taste.

129. *Zea mays* Linn.

Family: Poaceae

Common Name: Maize

Local Names – Chothe: *Chungachak,* **Manipuri:** *Chujak.*

Part Used: Fruit.

Diseases/Treatments: (i) Food poison, (ii) Wrong medication.

a. Crush the young maize into powder form (by adding little water) and immediately give the liquid paste form as much as possible to the patient. Give the solution several times even after the person vomited many times till it is fully cleanse and neutralised. (In certain case, when the fresh maize is not available, get the dry bud of the maize and crush it into powder form by adding little water and force the person to drink as much as possible till neutralised).

b. The same solution may be given to person in certain cases when a person took pesticides, insecticides, or wrong medicine or prescription.

How to Identify: The maize is cultivated by many people for home consumption or for economy. The plant grows upto a high of about 5-10 feet. The mature maize is golden yellow in colour wrapped with several layer of petals with long hair like structure dropping at the tip of the fruit.

130. *Zingiber officinale* Rosc.

Family: Zingiberaceae

Common Name: Ginger.

Local Names – Chothe: *Aithing*, **Manipuri:** *Shing.*

Part Used: Rhizome.

Diseases/Treatments: (i) Cough, Bronchitis, (ii) Body stimulant, (iii) Constipation, Piles, (iv) Extra Tissue growth.

a. When a person suffers from cough and bronchitis, one may chew pieces of fresh ginger time to time till one gets cured. About 5-10ml of the extracted liquid mixed with little salt may be taken before food till one gets cured.
b. Sometimes to enhance body stimulation, some pieces of fresh ginger may be added while boiling red tea and taken along with tea to enhance body stimulation or increase immune system.
c. When an adult suffers from constipation, he may insert a small amount of crushed fresh ginger mixed with little mustard oil or burnt kerosene residue inside his anus to remove the gas and old stole. Some believed that it even help from the minor piles problems if done occasionally.
d. Take a slice of fresh ginger and rub it against the extra tissue growth (of soft fleshy type) 2-3 times a day till cured.

How to Identify: There are varieties of ginger, local indigenous ginger, hybrid and common ginger. The Chothe besides using fresh ginger in curry also used it for certain ritual purposes especially in divination.

Chapter 3

Data Analysis of Chothe Ethmomedicinal Plants

Table 1: Chothe Indigenous Medicinal Plants and its Treatment Methods

Sl. No.	*Botanical Name [Family]- Common Name*	*Chothe Name- [H, Sh, T, C]*	*Manipuri Name [S/P]*	*Parts Used [Forms]- [O/E]*	*Diseases – Treatment Methods and its Dosages*
1	Acalypha hispida Burm. F. [Euphorbiaceae], Chenille plant, Red hot cat's tail.	*Kaprang arapa* [Sh]	*Kaprang manbi* [S]	Leaf [E]	White discharge of woman: - Crush the fresh leaves and add little limestone and apply on the body parts twice a day for a week or till cured.
2	*Acemella oleracea* (L.) R.K. Jansen [Asteraceae] Toothache plant, Sichuan buttons	*Ansa* [H]	*Morok sabi* [P]	Leaf/Flower [O]	i) Boil, ii) Urinary problem, iii) Toothache, iv) Worms (esp. child): i) The matured leaf is heated on fire and applied on the boil part, time after time till it is cured. ii) Those suffering from urinary disorder may eat fresh tender leaves as simple boiled curry.

Sl. No.	*Botanical Name [Family]- Common Name*	*Chothe Name- [H, Sh, T, C]*	*Manipuri Name [S/P]*	*Parts Used [Forms]- [O/E]*	*Diseases – Treatment Methods and its Dosages*
					iii) When a person suffers from toothache the flower bud may be crushed with hands and inserted in the cavity of the toothache or put around the aching tooth time after time till it gets relief. It acts as sedative. iv) Boil the leaves and give bath with the warm water to the baby 2-3 times occasionally once a month only.
3	*Achyranthes aspera* Linn. [Amaranthaceae], Devils horsewhip, Prickly chaff flower, Apang.	*Haotong koite* [H]	*Haodong thangol* [S]	Leaf [O]	Diarrhoea (Esp. for children): Collect some fresh leaves and cooked it as simple boiled curry along with dried fishes and serve 3 times a day for 3 days. This is especially for children above 2 years. The simple curry should be taken by the feeding mother too daily till cured.
4	*Acorus calamus* Linn. [Acoraceae], Sweet flag, Beewort.	*Oak lou* [H]	*Oak hidak* [S]	Rhizome. [E/O]	Worms/Ascaris (esp. for children): A piece of the rhizome is worn around the child's neck as necklace. Crush little of the rhizome and mixed little pure water and give a tea spoonful (2-3ml) twice a day only. Do not give excess to children.
5	*Adhatoda vasica* Nees. [Acanthaceae], Malabar nut.	*Annha angoupa* [Sh]	*Nongmangkha Angouba* [P]	Leaf. [O]	Jaundice: Cook some fresh tender leaves as simple curry along with dried fishes and have twice a day morning and evening for a week. It is strongly suggested to avoid red meat, chilly, fresh fishes). The taste will be little sour.
6	*Alangium chinense* (Lour.) Harms. [Cornaceae], Chinese langium.	*Kokal* [Sh]	*Kokal* [S]	Leaf. [O]	Fever: Crush the fresh leaves and take about 10-20ml of the extracted liquid 2 times a day till cured. For children the dosage is given about 2-5ml only. It tastes bitter.

Sl. No.	Botanical Name [Family]- Common Name	Chothe Name- [H, Sh, T, C]	Manipuri Name [S/P]	Parts Used [Forms]- [O/E]	Diseases – Treatment Methods and its Dosages
7	*Aleo vera* (L.) Burm.f. [Asphodelaceae], Aleo vera.	*Kheva arikpa/Aleo vera* [H]	*Aleo vera* [P]	Leaf [E/O]	i) Burn, ii) Chronic gastritis/ulcer, iii) Dark pigmentation: Immediately apply the *aleo vera* leave gel smoothly over the burn areas till the burning sensation is cold. This will reduce the watering of the skin and burnt marked. Eat the leave gel a spoonful twice a day in empty stomach and drink plenty of water after 15 minutes gap for a week or two weeks to help cure from chronic ulcer and gastritis. To cure the dark pigmentation on the face, apply the leave gel over the dark pigmentation over the face in the morning and evening for a month and also eat a spoonful of the fresh leave gel for two weeks.
8	*Allium sativum* Linn. [Alliaceae]- Garlic	*Satun* [H]	*Chanam* [S]	Bulb [Raw] – [O]	Fever, Paralysis, Muscular Pain: When a person suffers from slight fever, rheumatism, arthritis or semi-paralysis may apply the ointment prepared by adding few pods of garlic on the heated some pure mustard oil in a bowl. The fresh garlic bulbs are considered better. This ointment may be applied as balm and also used as massaging ointment in the painful joint areas, forehead, near the nostril, throat cleavage, chest, back, palm and feet till the body temperature returns to normal condition. This massaging if repeated regularly twice a day for a week or more to help reduce fever, rheumatism, arthritis, and muscular joint pains. Apart serving as daily culinary item the Chothes believed that the smell can dispel the evil spirit. Therefore, when they suffered from bad dreams or nightmares they used to keep it under their pillow or mattress.
9	*Alocasia indica* (Roxb.) Schott. [Araceae], Great-leaved Calendium, Giant Taro	*Jaitim* [H]	*Yendem* [P]	Stem/Leaf [O/E]	(i) Enhance blood (Post-natal woman), (ii) Boil: i) Traditionally the stem after removing the fibre or outer covering is cooked in a simple boil curry form along with fresh local fishes (Ngara) and given to post-natal woman to enhance the blood and enhance the stimulation. ii) Smear a piece of the leaf on the palm and add little saliva and apply the leaf on the boil area with little opening on the boil mouth. Repeat it 4-5 times after an interval of 6-8 hours till the pus is mature to burst.

Sl. No.	*Botanical Name [Family]- Common Name*	*Chothe Name- [H, Sh, T, C]*	*Manipuri Name [S/P]*	*Parts Used [Forms]- [O/E]*	*Diseases – Treatment Methods and its Dosages*
10	*Alpinia caerulea* (R.Br.) Benth. [Zingiberaceae] Native ginger	*Phuiphung* [H]	*Pulei manbi* [S]	Root [Dried] –[O]	Asthma and Tonsillitis: Person suffering from asthma and tonsillitis may drink about 10-20ml of the extracted liquid obtained from crushing the roots by adding little salt twice a day till it gets cured the tonsillitis. But for asthma patient one may take about 10ml time to time when required.
11	*Alpinia galangal* Linn. Willd. [Zingiberaceae]- Greater galangal	*Marou (Brou)* [H]	*Kanghu* [S]	Rhizome [Dry] – [O]	i) Piles, ii) Termination of Conception of a child (Abortion): i) Those persons suffering from piles may drink about 30-40ml of the extracted liquid obtained by crushing the rhizome in the morning for a week or till one is cured. For taste, a spoonful of honey may be added. ii) Woman who wants to terminate early conception of a child may drink about 50-60ml of the extracted liquid early in the morning in empty stomach till it ovulates.
12	*Amaranthus retroflexus* Linn. [Amaranthaceae] Spiny amaranth/ Spiny pigweed	*Si-ake anbu Arintoipa* [H]	*Chengkruk (Tingkhan-gpanbi)* [S]	Leaf, Stem [Raw] – [E]	Extra tissue growth, Haemorrhoid: Apply little lime (limestone) on the mother (first) extra tissue growth or haemorrhoid whether on the hand or leg. Then, when the tissue becomes little soft smeared with the matured stem (stalk) repeatedly against the extra tissue growth by pouring water above it. The process of applying lime, smearing it with the stalk of *Amaranthus spinosus* and washing with water is repeated till it bleeds. It is believed that if the mother tissue cell is completely uprooted the other tissues would miraculously fade away slowly. The Manipuris called it *sarik* because it grows in branches. It is not Keloid but looks alike.
13	*Anisomeles indica* (L.) Kuntze. [Lamiaceae], Catmint	*Sipeng arapa* [H]	*Thoiding angouba* [P]	Leaf [O]	Piles: Boil some leaves and served half a glass twice a day for 3-5 days or till cured of the piles.

Sl. No.	*Botanical Name [Family]- Common Name*	*Chothe Name- [H, Sh, T, C]*	*Manipuri Name [S/P]*	*Parts Used [Forms]- [O/E]*	*Diseases – Treatment Methods and its Dosages*
14	*Antidesma acidum* Retz. [Phyllanthaceae] Amita	*Tuitrit* [Sh]	*Ching Yensin* [P]	Leaf [Raw] – [O]	Indigestion and Enhance Immune System: Person suffering from indigestion or wish to enhance immune system may drink half a glass of the boiled leaves as decoction anytime of the day (better after food), this enhanced the blood circulation and increase the immune system and indigestion.
15	*Arisaema tortuosum* Schott. [Araceae], Whipcord cobra lily, Bagh jandhra	*Arui cheisu* [H]	*Lin cheisu* [S]	Rhizome [O]	Virus infections/Skinny, Pricked by poisonous plants like boils: Boil the fresh rhizome for a long time and let it cool. Then crushed the rhizome and bandage the paste around the infected wound area by the virus or bacteria. [Infection such as *Hin-naba, Lainingthougi tenna naba (lit. god's arrow), Lambi tingkhang (lit. thorn of footpath), Lai matu* (lit. gods hair) and *etc.* are treated with this herb].
16	*Artemisia vulgaris* Linn. [Asteraceae], Indian worm wood, Sagebrush	*Leikha lei* [H]	*Leibakngou akuppa* [P]	Leaf [O]	Diabetes: Boil the leaves and add little *sitamasi* and served half a glass of the decoction twice a day for 7 days or till cured.
17	*Averrhoa carambola* Linn. [Oxalidaceae], Star fruit	*Heinouchom* [T]	*Heinoujom* [S]	Fruit [E]	(i) Mud stain on nails, (ii) Germicides: i) The fruits or the extracted liquid is use to remove mud stain on finger nails by polishing with the fresh fruits especially by the local paddy transplantation during the season. ii) The fruit are crush and thrown into the polluted water and pond banks of fishery farm affected from severe flood to kill the germs and bacteria that produces very awful stinky smell.
18	*Azadirachta indica* A.Juss. [Meliaceae]- Neem	*Theichak* [T]	*Neem* [S]	Leaf and bark [Raw]-[E]	i) Cough, Skin diseases, ii) Toothache: i) Person suffering from cough and other skin diseases may boiled the fresh leaves in a pot of water and bath everyday till it gets cured. ii) The bark or twig of Neem is used in cleaning the teeth or for toothache in the past.

Sl. No.	*Botanical Name [Family]- Common Name*	*Chothe Name- [H, Sh, T, C]*	*Manipuri Name [S/P]*	*Parts Used [Forms]- [O/E]*	*Diseases – Treatment Methods and its Dosages*
19	*Basella alba* Linn. [Basellaceae], Malabar spinach, Vine spinach, Buffalo spinach.	*Noinotte* [C]	*Urok sumban* [P]	Leaf [E]	Boil: Heat abit on the fire the fresh leaves and apply the soft leaves on the boil area with little opening at the tip of the boil. Repeat this process till the boil become ripe and about to burst out the pus. It is suggested that one should not mess with the boil until it is fully ripe for it becomes later cancerous and sceptic.
20	*Bauhinia variegate* (L.) Benth. [Fabaceae] Orchid tree, Mountain-ebony	*Chingthrao/ Heina kundo* [T]	*Chingthrao/ Heina kundo* [S]	Leaf [O]	(i)Anaemia Blood purification, (ii) Reduce weight: i) Cooked some leaves as a simple curry along with fresh small shinny local fishes or prawn (*Ngakha* or *Khajing*) and eat twice a day (Morning and evening) for atleast 5 days to improve once health from weak and sickly conditions. ii) Perhaps, if the fresh leaves are cook as simple curry with dried small shinny local fishes (Ngakha) twice a day for 5 days or more, the person's weight gradually loses or decreases.
21	*Benincasa hispida* Thunb. Cogn. [Cucurbitaceae]- Ash-gourd/Winter melon	*Maipoy/ Maingou* [C]	*Torbot* [S]	Fruit [Raw] – [O/E]	i) Giddiness, Nausea, ii) Chronic-ulcer: i) Person suffering from frequent giddiness or nausea may apply the paste of this matured *Benincasa hispida* fruit above the forehead to sooth and calm down the blood pressure. The preparation method is that the outer green cover of the fruit is remove and the inner white fruit is sliced into small pieces (semi-paste form) with a spoon without touching one's hand for the required purpose. The paste is poured into a thin cloth and wrapped directly above the forehead. ii) When a person suffers from chronic-ulcer he may either eat fresh in raw form in empty stomach each morning and evening or boiled and have it as simple cooked curry regularly till cured. It serves as anti-oxidant.
22	*Bischofia javanica* Blume. [Phyllanthaceae], Bishop wood.	*Uthum toh* [T]	*Uthum panbi* [S]	Leaf (Tender). [O]	High Blood pressure: The tender fresh leave are either boil and eaten as simple boiled or the fresh leave is eaten along with local chutney along with food occasionally as food items to reduce the blood pressure.

Sl. No.	*Botanical Name [Family]- Common Name*	*Chothe Name- [H, Sh, T, C]*	*Manipuri Name [S/P]*	*Parts Used [Forms]- [O/E]*	*Diseases – Treatment Methods and its Dosages*
23	*Brassaiopsis hainla* (Buch.-Ham.) Seem. [Araliaceae], Chuletro, Seto Chuletro Tree	*Chom* [Sh]	*Chom (variety)* [S]	Leaf [O]	Kidney failure, Urinary infectious problems: Boiled the fresh leaves in about 2-3 litres of water till it turns deep red colour and one may serve a glassful of the decoction twice a day for a week. In case of improvement the patient may continue to take the decoction solution till fully cured for a month. During the medication avoid any fresh red meat, oils, and spicy things. One may also crush the leaves and take half a glass of the extracted liquid before food once a day for 3 days. If there is a sign of improvement continue for a week or till fully cured.
24	*Cajanus cajan* (Linn.) Millsp. [Fabaceae], Pigeon pea, Red gram.	*Meirongpi* [Sh]	*Meirongbi* [S]	Leaf [O]	Dysentery: Crush some fresh leaves and add little common salt and take about 20-30ml 2/3 times a day for 2-3 days.
25	*Cannabis sativa.* [Cannabaceae] True hemp	*Ganja* [H]	*Ganja* [S]	Leaf [Raw]- [O]	i) Diarrhoea, Dysentery, ii) Lack of appetite, iii) Depression: i) Person suffering from diarrhea or dysentery may drink about 10-20ml of the extracted liquid obtained by crushing the leaves once a day after food till cured. ii) Person suffering from lack of appetite may fry few leaves mixed with some flour and taken some before food. iii) Person suffering from serious depression may smoke mildly the dried leaves in pipes or as cigarette. (It is suggested only if the person is completely free and willing to rest for few days from serious works).
26	*Capsicum annum* Var. fasiculatum Baily. [Solanaceae], Bhut jolokia/Red Naga chilly	*Masi arokpa/ U-morok* [H]	*U-Morok* [S]	Fruit [E]	Ringworm: Cut fresh ripe Naga chilly into halves and apply on the ringworm area. Repeat this 3-4 times alternate days till it is cured.

Sl. No.	*Botanical Name [Family]- Common Name*	*Chothe Name- [H, Sh, T, C]*	*Manipuri Name [S/P]*	*Parts Used [Forms]- [O/E]*	*Diseases – Treatment Methods and its Dosages*
27	*Carica papaya* L. [Caricaceae]- Papaya	*Awath* *pi* [Sh]	*Awathabi* [S]	Fruit [Raw] – [O/E],	i) Chronic-ulcer, Gastritis, ii) Dog bites: i) Person suffering from severe or mild chronic-ulcer may eat lots of papaya either ripe or unripe daily till cured. ii) When a person is bitten by a dog immediately apply a slice of the green outer cover along with the pus on the bitten spot in order to neutralize or extract the poison and one may eat lots of unripe papaya too.
28	*Cassia timoriensis* Wlld. [Caesalpiniaceae] Stout prickly climber	*Laikee* [Sh]	*Laikee* [S]	Stem (Core part) [O]	Jaundice, Kidney failure: Boil for long time few small sliced of the dried stem (core part of the base stem) and when it turns blood red, cold it down and take 20-30ml of the decoction solution twice a day for 5 days after food. If there is a sign of improvement one may continue to take the prepared solution once every day and when required till fully cured. For children suffering with jaundice below the age of 2-3 months old give 2-5 drops of the solution once a day only before feeding till cure.
29	*Celtis timorensis* Linn. [Ulmaceae], Southern nettle.	Heikreng [T]	*Heikreng/ Heigreng* [S]	Leaf/Bark [O]	Stone cases (Kidney): Boil the fresh leaves and bark and have the decoction 3 times a day till cured.
30	*Centella asiatica* Linnaeus. Urb. [Apiaceae]- Asiatic/Indian penny wort	*Anleiphon/ Ariphon-* [H]	*Peruk* [S]	Leaf and stem (whole plant) [Raw]- [O]	i) Hypertension, Gastritis, Chronic ulcer, ii) i) Constipation, ii) Blood Purification: i) Persons suffering from serious hypertension, gastritis, chronic ulcer may take about 30-60ml of the extract liquid obtained from crushing the whole plant (leaf and stem). A spoonful of honey may be added or taken singularly as tonic before each meal regularly for a period of 1 week or more, depending upon the seriousness. If needed, the liquid may be diluted with some water and taken regularly. ii) Persons suffering from weak immunity, constipation or wished to enhance blood purification system may take a glass full of the boiled plant as concoction after or before meal or anytime for a week or more.

Sl. No.	*Botanical Name [Family]- Common Name*	*Chothe Name- [H, Sh, T, C]*	*Manipuri Name [S/P]*	*Parts Used [Forms]- [O/E]*	*Diseases – Treatment Methods and its Dosages*
31	*Cinnamomum tamala* (Buch-Ham) T.Nees. [Lauraceae], Indian bay leaf/Tajbat	*Tastepat* [T]	*Tastepat* [S]	Leaf [O]	Urinary infectious problem: Boil some fresh tastepata leaves till it turns red and add little *Sitamasi* (crystal cane sugar) and serve about half a glass or glassful of the decoction solution once a day for 3 days or till cured. (It is suggested to take immediately when the irritation just began. One may also take the solution instead of water when demanded).
32	*Cissus adnata* Roxb. [Vitaceae]	*Kokngoujen* [Cl]	*Kokngouyen angouba* [S]	Leaf [O]	Stone cases (Bile/Kidney/Urinary): Boiled some fresh leaves of *Cissus adnata* along with some piece of Upupa epops (Hoopoes) bird's meat and served it as stew or simple boiled curry for a week to cure from any kidney/liver/bile's stone cases.
33	*Citrus × aurantifolia* Linn. Burm [Rutaceae], Common/Sour Lime	*Champra* [Sh]	*Chambra* [S]	Fruit [Fresh]- [O]	Kidney (Stone) problems: Person suffering from kidney stone problem may drink the mixture of fresh ripen extracted lemon juice with a local hen egg in empty stomach in the morning. The preparation method is that, the fowl's egg is dip into the glass half filled with the extracted lemon juice for about 12 hours. Then, the un-dissolved materials like the cellulose of the egg and bit of shell may be removed slowly and after which stirred well the mixture and drink it in empty stomach early in the morning. Later after 10-20 minutes, one may drink lots of water before anything is taken. This process may be repeated and taken till one gets relief of the pain/ stone. Preparation for this medicine required highly concentrated type of lemon which is locally grown in the hills which is very acidic and a local fowl egg. One should note that after about 6 hours if the egg shell does not dissolve when rotating the un-dissolved upper egg shell, it may not be used at all, signifying that the lemon is of low grade quality. Have the solution only if the egg shell fully dissolved after 6 hours. This process may be repeated if needed.
34	*Citrus maxima* (Burm.) Merr. [Rutaceae], Pomela/Citrus grandis	*Nobab* [Sh]	*Nobab* [S]	Fruit [E]	Conjunctivitis Apply 1-2 drops of the ripe fruit taken from the bigger extreme corner of the fruit capsule on the sore eyes or person easily infected with conjunctivitis.

Sl. No.	*Botanical Name [Family]- Common Name*	*Chothe Name- [H, Sh, T, C]*	*Manipuri Name [S/P]*	*Parts Used [Forms]- [O/E]*	*Diseases – Treatment Methods and its Dosages*
35	*Clerodendrum colebrookianum* Linn. Moon. [Lamiaceae]- Bleeding heart	*Anphui (Soknoudon)* [Sh]	*Moirang Khanum* [S]	Leaf [Raw] – [O],	Hypertension and Controls high Blood Pressure: Persons suffering from hypertension, anxiety and high Blood Pressure may take a glass full of the fresh boiled leaves of *Clerodendrum colebrookianum* after each meal or anytime for 2-3 days or till the pressure reduces. However, prolong or excess intake of this decoction is considered bad as it lowers the blood pressure again. Or the patient may consume some amount of the boiled leaves along with other food items occasionally.
36	*Costus speciosus* (J. Konig.) Smith. [Costaceae], Spiral ginger	*Oakchak khombi* [H]	*Oakchak khombi* [S]	Stem [E]	Conjunctivitis, Sore eyes, Ear liquid: Crushed the clean stem and add 2-3 drops into the sore eyes or ear infection side 2-3 times a day for 7 days or till it is cured.
37	*Cretaeva religiosa* G. Forster. [Capparidaceae], Temple plant, Sacred garlic pear	*Panthing lei* [T]	*Loiyumba mapal/Loiyum lei* [S]	Leaf/Root [O]	Joint pains: Boiled some amount of the fresh leaves and when cold, bandage the leaves around the joint pains like knee or hips, *etc.* Repeat this process after 8 hours or the next day till it is relief and cured.
38	*Curcuma caesia* Roxb. [Zingiberaceae] Black turmeric/Black zedoary	*Aisan ahangpa* [H]	*Yaimu* [S]	Rhizome [Raw] – [O]	i) Smallpox, Tumour, ii) Dispeller of Magical Spell: i) When a person suffers from smallpox or minor tumour the crushed rhizome is mixed with little mustard oil and is applied on the smallpox area or around the tumour. This procedure is repeated till the disease is cured. The patient (if feels) may even eat some amount of the fresh or dry rhizome about 10-20mg after food. ii) Traditionally, some local medicine man/priest asked a person to drink the extracted crushed liquid of this rhizome mixed with little water and salt after offering ritual when considered suffering from the magical spell and charms. This tonic is taken for 3-4 days. This *Curcuma caesia* rhizome (ginseng) is considered very rare to find.

Sl. No.	*Botanical Name [Family]- Common Name*	*Chothe Name- [H, Sh, T, C]*	*Manipuri Name [S/P]*	*Parts Used [Forms]- [O/E]*	*Diseases – Treatment Methods and its Dosages*
39	*Curcuma longa* Linn. [Zingiberaceae] Turmeric	*Aisan* [H]	*Yaingang* [S]	Rhizome [Raw] [O]	i) Carminative Stimulant, ii) Intestine Worms, iii) Cut, iv) Complexion: i) About 20-30ml of the extracted liquid obtained by crushing the raw rhizome of *Aisan* and by mixing with little water and salt may be taken by person suffering from indigestion, weak abdomen muscle bowel and carminative problems. This tonic should be taken before food in empty stomach atleast 3-4 times a day for a week. However, it is recommended to have once or twice in a month as health precaution since it helps in strengthening the bowel muscles and stimulates the digestive systems. ii) This turmeric tonic is also given to children believed to suffer from infested intestinal worms like tape worms, ascaris, *etc.* 2-3 times in a week. iii) When a person suffers from a major cut or wound on any body parts, the fresh turmeric is crush and the paste mixed with little mustard oil and lime (limestone) is applied on the cut area and bandaged for 2-3 days. Because of its special chemical property as a good agent of blood clotting and antiseptic, the wounds are healed faster with dry skin. iv) If the fresh turmeric is slightly roasted and eaten as chutney regularly it is considered to enhance one's complexion and free from pimples.
40	*Curcuma zedoaria* (Christm.) Roscoe. [Zingiberaceae] White turmeric, Kentjur.	*Aithing angoupa* [H]	*Sing angouba* [S]	Rhizome [O]	Avoid nerves constrain, Stomach ache (To ward off evil spirit): Chew a small piece of the rhizome early in the morning to relief from stomach ache and to ward off from evil eyes and evil spirits so that nothing happens to you like stomach ache or head ache on the journey.

Sl. No.	*Botanical Name [Family]- Common Name*	*Chothe Name- [H, Sh, T, C]*	*Manipuri Name [S/P]*	*Parts Used [Forms]- [O/E]*	*Diseases – Treatment Methods and its Dosages*
41	*Cuscuta reflexa* Roxb. [Convolvulaceae] Giant dodder.	*Jamrui anemte* [C]	*Kunjarata* [P]	Stem [O]	Diabetes (Urine case), (ii) Jaundice: i) Crushed a bit and boil some amount of this yellow creeper plant in about 2-3 litres of water till it turns deep brown. When it gets cold take a glassful three times a day after food for a month or so. It is suggested that the decoction solution should be taken instead of normal water. ii) Boil some of the creeper along with big water bug (*Esing Nausek*), ghun seed and a Longpal seed and serve a glassful of the decoction solution 2 times a day for 7 days or more. The solution can also be used for bathing, washing hands and face occasionally.
42	*Cynodon dactylon* Pers. L. [Poaceae] Doob/Bermuda-grass	*Tingthou/ Bangru-bange/ Sanahuplang* [H]	*Tingthou* [S]	Leaf [Raw] [O]	i) Urinary Problem, ii) Children stomach problem, iii) Cancer: i) If a person suffers from urinary problem, take about 2 tea spoonful of the crushed young tender leaves of Cynodon dactylon (Tingthou) mixed with little water and honey early in the morning before food till cured. ii) Some priests also suggested the extracted liquid of the crushed leaves and stems to children suffering from intestinal or stomach problems to give about 5-10ml twice a day for a week before meal. iii) Crushed some amount of the fresh leaves of tingthou and add about 2-5ml of the extracted ginger liquid and take about 10 ml twice a day for a month and if helpful one may continue. Boiled a handful leaves of the Cynodon dactylon and Polygonum odoratum (Phakphai) in 5 litres of water and collect only about 1 litre and served the decoction half a glass twice a day before meal till pressure reduced.

Sl. No.	*Botanical Name [Family]- Common Name*	*Chothe Name- [H, Sh, T, C]*	*Manipuri Name [S/P]*	*Parts Used [Forms]- [O/E]*	*Diseases – Treatment Methods and its Dosages*
43	*Dichrocephala integrifolia* (L.f) Kuntze. [Asteraceae] Dicrocefala in Italian, Phakchee doi in Thailand.	*Aralu* [H]	*Lalukok* [S]	Whole plant [O]	(i) Enhance blood (Post-natal woman), (ii) Shampoo: i) Cooked the tender shoots/leaves as simple boiled curry along with fresh local fishes (*Ngara*) and serve occasionally for a week or two. This is highly recommended for the post-natal woman to enhance their blood considered to be rich in iron and other minerals. ii) Some fresh leaves are added in local made shampoo of rice starch and orange cover. It is considered that when this local shampoo is used regularly it keeps the head cold. The hair also becomes black, thick, smooth and shiny.
44	*Dipterocarpus turberculatus* [Dipterocarpaceae] Gurjan, Teak	*Khangra* [T]	*Khangra* [S]	Leaf [O]	Menstrual problems: The tender leaves are boiled and served as simple cooked curry with dried fishes for 3-4 day or occasionally till cured.
45	*Drymaria cordata* (L.) Willd. [Caryophyllaceae] Tropical chickweed	*Tandan tadit* [H]	*Tandan tadit* [S]	Whole plant [O]	Typhoid, Internal inflammation of stomach, Crack of lips: Crush the whole plant and add little Sitamasi (crystal cane sugar) and take half a glass twice a day for 2-3 days before food especially, early in the morning.
46	*Dysoxylum densiflorum* (Blume) Mig. [Meliaceae] Indian white cedar	*Mishribu* [T]	-- [P]	Leaf [E]	Cancer, Tumour: Crush the fresh leaves and apply about 5 ml of the extracted liquid solution on the affected cancer or tumour area once a day till cure.
47	*Eclipta alba* (Linn.) Hassk. [Asteraceae], Trailing eclipta	*Uchi sumban* [H]	*Uchi sumban* [S]	Leaf [E]	Inflammation of stomach, Crack on the lips and tongue: Crush the fresh leaves and apply the extracted liquid time to time on the crack lips and tongue.
48	*Elatostema lineolatum* Wight. [Urticaceae], Lined Elatostema, Ediew	*Tuidaitoh* [H]	[S]	Leaf [O]	Gastritis, Ulcer, Liver: Boil the tender leaves for about 5 minutes and have the boiled leaves in empty stomach and one may also have the decoction twice a day for 5-7 days to reduce from gastritis, ulcer and liver problems as it cools down the stomach from acidity. One may also cook along with rice and meat and have as stew regularly or occasionally to reduce acidity.

Sl. No.	*Botanical Name [Family]- Common Name*	*Chothe Name- [H, Sh, T, C]*	*Manipuri Name [S/P]*	*Parts Used [Forms]- [O/E]*	*Diseases – Treatment Methods and its Dosages*
49	*Elsholtzia griffithii* [Lamiaceae] Lengser Rep, Lengmesel	*Lengtu* [H]	*Lomba* [S]	Leaf and flowers [Raw] [O]	Asthma, Nervous spasms: Person suffering from asthma and nervous spasms may take about 2 tea spoonful of the extracted liquid obtain by crushing the leaves and adding some water twice a day before each meal, till it gets cured. The leaves are elongated and coarse. The greyish flower is sponge and seed are tiny. It grows about 3-5 feet in high. This *Elsholtzia griffithii* plant is cultivated in kitchen garden by most tribal communities. The flowers are preserved and added as flavour in chutney, fish, beef and yam curries as delicacies.
50	*Elsholtzia ciliata* (Thunb.) Hyland. [Lamiaceae] Vietnamese balm	*Areng* [H]	*Teckta* [S]	Leaf and flower [O]	(i) Worm (babies), (ii) Bad old gas of stomach, (iii) Cough and cold fever: i) Smear some fresh leaves with your palm and add little common salt and pure oil and apply the paste solution as balm on the baby's body (except the stomach area) to children infested with intestinal worms. ii) Chew some fresh leaves and drink some water 2 times a day for 3 days or more. iii) Smear some fresh leaves on your palm and add little common salt and pure mustard/olive oil and massage the whole body with the ointment as balm especially the feet, chest, hands, back and around the nostril area. But it is strongly suggest avoiding the massage around the stomach and belly areas.
51	*Entada phaseoloides* (L.) Merr. [Fabaceae] St. Thomas bean	*Kaang* [Cl]	*Kangkhil (Kang)* [S]	Seed [E]	Scare mark, Bruise: Keep smearing slightly the dried rounded seed 2-3 times a day to eradicate the bruise or scare mark by spelling or saying subside and get rid of the scare mark 3-5 times.
52	*Eryngium foetidum* Linn. [Apiaceae] Saw tooth	*Awa-neem (Somey)* [H]	*Awa-phadigom* [S]	Leaf or whole plant [Raw] – [O]	Liver and Hepatic problems: Person suffering from liver and hepatic problems may take about 10-15ml of the extracted liquid of the leaves obtained by crushing and adding a tea spoonful of honey regularly before each meal for a period of 1- 2 weeks.

Sl. No.	*Botanical Name [Family]- Common Name*	*Chothe Name- [H, Sh, T, C]*	*Manipuri Name [S/P]*	*Parts Used [Forms]- [O/E]*	*Diseases – Treatment Methods and its Dosages*
53	*Erythrina variegate (alba)* Linn. [Fabaceae] Indian coral tree, Lenten tree, Tiger's claw.	*Hungthu angoupa* [T]	*Kurao angouba* [P]	Bark [E]	(i) Dispeller of evil spirit, (ii) Possess by evil spirit: i) Cut a piece of the bark and polished it well into small pieces and is often wear around the neck of a child or on the arm band or waist of an adult to ward off evil spirits by most locals including the Meitei's (Manipuris). ii) Cut one of a big branch or stem and make a drum shell when dried by curving out a hollow, and cover and tide the two openings with the vulture's skin/hide. Perhaps, when the drum is beaten the evil spirit comes out and dances to the tune. (This is a belief that may perhaps involve magico-religious ritual).
54	*Eupatorium cammanoi* Linn. [Asteraceae] Hempagrimony	*Aripung/ Renglei* [H]	*Langthrei* [S]	Leaf [Raw] – [O],	i) Acidity, Gastritis, ii) Burning Sensation: i) Person suffering from frequent acidity or gastritis/gas formation in the stomach may take about 10-20ml of the extracted liquid of the leaves obtained by crushing in empty stomach before each meal or atleast three times in a day. For children one may dilute the liquid with little water in a glass mixed with little salt and have it. ii) One may eat few fresh leaves and drink a glass of water if a person suddenly suffers from burning sensation of stomach from excess of eating chilly or sudden development burning sensation. In certain cases, one may have it in tonic form as described above for immediate relief from the pain.
55	*Ficus hispida* Linn. f [Moraceae] Hairy fig/Devil's fig	*Theichang* [T]	*Heirik* [S]	Leaf [Raw]- [E]	Ringworm and Scabies: Person suffering from ringworm may rub the leaf against the ringworm infected area or on the scabies part still it bleeds little. Later one may wash it with clean water and apply some pure (mustard) oil around it. This may be repeated 3-4 times till cured. The leaves are broad, big and very coarse. The fruits are little rough, rounded in structure. Green and brownish when unripe and turns red when ripe. It bears in cluster from a string of root like stalk usually at stem of the base.

Sl. No.	*Botanical Name [Family]- Common Name*	*Chothe Name- [H, Sh, T, C]*	*Manipuri Name [S/P]*	*Parts Used [Forms]- [O/E]*	*Diseases – Treatment Methods and its Dosages*
56	*Ficus racemosa* Linn. [Moraceae]	*Athi theichang* [T]	*Asi Heibung* [P]	Leaf, Fruit [O]	Diabetes: Chew some fresh tender leaves and young fruits of *Ficus racemosa* 2 times a day for 1-2 months regularly.
57	*Ficus tsjahela* Burm. Hook.f. [Moraceae] Pilkhan, Karal fig, South Indian fig.	*Abung* [T]	*Khongnang Tarung* [S]	Leaf [O]	Diabetes: Boiled some leaves in about 3 litres of water and have a glassful of the decoction twice a day regularly and occasionally till reduced of the blood sugar level.
58	*Gomphrena celosioides* Mart. [Amaranthaceae] Bachelor's button, Prostrate Globe-Amaranth	*Noonte arapa* [H]	*Chini kang* [S]	Whole plant [O]	Stone cases: Boiled the whole plant of *Gomphrena celosioides* in about 2 litres of water and served the decoction solution a glassful twice a day for 5-7 days or till relief.
59	*Goniothalamus sesquipedalis* (Wall.) Hook. F. & Thomson. [Annonaceae] Goniothalamus	*Aham* [S]	*Leikham* [S]	Leaf [E]	Fever, Stomach ache (esp. children): A piece of the dried leaf is burnt, and the ash collected in a glass and little water is added. Drink little amount of the mixed solution and the rest is applied on the whole-body parts. This is usually done twice or so only when a child suffers from sudden fever or stomach upset with a belief of contact of evil spirits. If the fever does not subside, then, it is considered a normal fever and other medication are given. The dried leaves are burnt to dispel evil spirits in the house.
60	*Gynura cusimbua* D. Don/S. Moore. [Asteraceae] Hill gynura	*Muhun lou* [H]	*Tera paibi* [S]	Leaf [Raw] – [E]	Cuts and Wounds (ii) Local shampoo (*Chinghe*): i) When a person suffers from cut and wound apply immediately the crushed leave paste to stop from bleeding and help blood clotting. The plant is also acts as antiseptic. ii) The fresh leaves of *Gynura cusimbua* and pumpkin tender leaves (*Amai bu/Maireng mana)* are added in the rice starch and left it to soak for 3-7 days after which the liquid is used to wash the hair to keep the head cool and hair smooth and silky.

Sl. No.	*Botanical Name [Family]- Common Name*	*Chothe Name- [H, Sh, T, C]*	*Manipuri Name [S/P]*	*Parts Used [Forms]- [O/E]*	*Diseases – Treatment Methods and its Dosages*
61	*Hedychium coronarium* (J.Koening). [Zingiberaceae] White garland lily, White ginger lily. (*Zingiber cassumunar* Roxb.)	*Aidou* [H]	*Yaingang ariba* [S]	Rhizome [O]	i) Irregular menstrual problem, (ii) Gastritis or joints and muscle pain due to internal bad gases (*Dadhu nungsit tharoupa*): i) Crushed the rhizome and add little common salt and take the solution of the extracted liquid about 5-10ml twice a day for 5 days before food in empty stomach early in the morning. ii) Boil the compound items of some *Aidou* rhizome, ghun seeds, Longpal seed, Jaiphor seed, garlic, local onion buds, a crab and white buffalo fats and take about 20-30ml twice a day for 3-5 days. One may also apply the solution on the joint pain areas 3-4 times a day for 3-5 days.
62	*Hedychium flavum* Roxb. [Zingiberaceae] Ginger lily	*Soklei/Iphung* [H]	*Loklei* [S]	Stem (young shoot) [O]	Irregular menstrual problem: Boil the young shoots in some amount of water and take half a glass twice a day for a week.
63	*Hedyotis auricularia* Linn. [Rubiaceae] Gatekola	*Langpan Koukha* [H]	*Langban Koukha* [S]	Stem, Leaf [O]	White discharges of woman: Boiled the leaves and stem and take a glassful of the cold decoction solution 3 times a day for 5 days in empty stomach.
64	*Hibiscus acetosella Welw. ex.Hiern.* [Malvaceae] Cranberry hibiscus	*Anthui ashangpa* [Sh]	*Sougree anganba* [S]	Leaf [E]	Boil, Boil on the back (*Khoi mahum* in Meitei): Boil the fresh leaves till it turn into paste form, then apply the paste over the infected boil area on the back side of the body. Crushed the leaves and make into paste and apply on the boil areas.
65	*Hibiscus cannabinus* Linn. [Malvaceae] Green/Red sorrel	*Anthui* [Sh]	*Sougree (ashinba)* [S]	Leaf and seed cover [Raw/dried] – [O]	Jaundice, Enhance Immune Systems and Refreshes the body: Person suffering from jaundice may boil the fresh leaves, or along with the soft outer covering of the seeds in about 1 litre of water and take half a glass as decoction twice a day anytime. Even the cooked leaves maybe eaten. There are two common types found in Manipur *i.e.* white and red species. The flower is big, white and dark purple in the centre. Occasionally eating fresh leaves enhanced body stimulation or immune systems.

Sl. No.	*Botanical Name [Family]- Common Name*	*Chothe Name- [H, Sh, T, C]*	*Manipuri Name [S/P]*	*Parts Used [Forms]- [O/E]*	*Diseases – Treatment Methods and its Dosages*
66	*Hibiscus rosa-sinensis* Linn. [Malvaceae], China rose,	*Juba kushum* [Sh]	*Juba kushum* [P]	Root [O]	Irregular menstrual cycle of woman: Boil some of the roots and take a glassful 2 times a day till cure.
67	*Holskioldia sanguine* Retz. [Lamiaceae] Chinese hat plant, Cup and Saucer	*Leiningthung asanpa* [Cl]	*Kharam leisok anganba* [S]	Leaf [O]	Boils, Body swelling because of bacterial infection, Scabies: Boiled the fresh leaves of *Holskioldia sanguine* for a while and bandage the leaves around the swollen infected area 2 times a day till cure. It is suggested to avoid eating any fermented things like local dry fishes (*Ngari, handak, hawajar*) during medication.
68	*Houttuynia cordata* Thunb. [Saururaceae]- Lizard tail, Heartleaf	*Anleithaang* [H]	*Toningkok* [P]	Leaf and stem [Raw] – [O],	Gonorrhoea: Those persons suffering from gonorrhoea may take about 20-30ml of the extracted liquid obtained by crushing the leaves and stems and adding little common salt before each meal twice a day for a week or till one gets cured.
69	*Hydrocotyle sibthorpioides* Lam. [Araliaceae], Khulkhuri.	*Lai peruk* [H]	*Lai peruk* [S]	Leaf [O]	Stone cases: Crush the fresh leaves and add little water and take about half a glass twice a day for 5-7 days. (One is suggested to drink plenty of water during this medication).
70	*Impatients balsamina* Linn. [Balsaminaceae], Garden balsam, Rose balsam.	*Khutshang* [H]	*Khutjang* [S]	Leaf [E]	Cuts and wounds, Bruise, Wart: Crush the fresh leaves and immediately bandage it the paste on the cuts and wounds especially, the bruise finger nail when hit by a hammer or hard substance.
71	*Ipomoea batatas* (Linn.) Lam. [Convolvulaceae] Sweet potato	*Alu Mangkara* [C]	*Mangara (Alu mangara)* [S]	Leaf [E]	Giddiness/Migraines (esp. woman): Crushed some fresh leaves and add little pure oil like Same, olive oil, and apply the paste above the forehead. This will shoot the head by keeping it cold thereby reducing the giddiness or dizziness or headache or migraine often occurred in elderly woman.

Sl. No.	*Botanical Name [Family]- Common Name*	*Chothe Name- [H, Sh, T, C]*	*Manipuri Name [S/P]*	*Parts Used [Forms]- [O/E]*	*Diseases – Treatment Methods and its Dosages*
72	*Ipomoea carnea* Jace. [Convolvulaceae] Morning glory tree	Lamkhop lei [Sh]	Sambal-pambi/ Lamkhop lei [P]	Leaf [O]	Menstrual problem of white discharge: Boil some of the fresh leaves along with along with the whole plant of *Kangphang ikaithabi* (*Mimosa pudica* Linn.) by adding little amount of sitamsi (crystal cane sugar) in about 2-3 litres of water and drink a glassful of the decoction solution twice a day in the morning and evening before food for a week or more. For those who are sensitive to the solution they got cured. (One should exclude the flower in the preparation).
73	*Isodon ternifolius* D.Don. [Lamiaceae] Kudo	*Koichu* [Sh]	*Khoiju* [S]	Leaf [E]	Fever, Ward of evil spirit: The dried leaves are burnt as incense to ward of evil spirits when a person suddenly falls sick. The ashes are also applied on the forehead and body parts.
74	*Jatropha curcas* Linn. [Euphorbiaceae] Physic nut	*Awa keke* [Sh]	*Awa keke* [P]	Leaf/Stem Gum [E]	Cut and wounds: Apply either the leaf or stem gum on the minor cut or wound time to time till the blood stops. It acts as first aids medicine.
75	*Justicia adhatoda* Linn. [Acanthaceae] Adulsa/Malabar nut	*Anha (Chikpa)* [Sh]	*Nongmankha* [P]	Leaf and flowers [Raw] [O]	i) Cough and Cold, Rheumatism, Feverish, Muscular/Joint pains, ii) Termination of Conception: i) For those who are suffering from slight cough and cold, and muscular/joint pains some leaves are added in hot boiling water and is taken down after few minutes. Served half a glass the decoction (or add little common salt) twice a day after each meal or atleast three times a day till cured. ii) About 50-60ml of the extracted liquid from the fresh leaves is drunk early in the morning in empty stomach by a girl or woman who wants to terminate early conception of a child or for abortion. If required they may also have in the evening when the stomach is empty, for a period of 1 week or more. (This is considered a secret treatment).

Sl. No.	*Botanical Name [Family]- Common Name*	*Chothe Name- [H, Sh, T, C]*	*Manipuri Name [S/P]*	*Parts Used [Forms]- [O/E]*	*Diseases – Treatment Methods and its Dosages*
76	*Kaempferia rotunda* Linn. [Zingiberaceae] Indian crocus.	*Leipak lei* [H]	*Leibak lei* [S]	Leaf [E]	Hot head, high Blood pressure: Boil the fresh leaves along with the local shampoo (*Chinghe)* and wash the head occasionally or alternately to relief from hot head and high blood pressure.
77	*Laggera alata* (D.Don). [Asteraceae] Oliver	*Phephe* [H]	*Haochak* [S]	Leaf [E]	Cough (esp. for children): Collect few good leaves and heat near the fire and when it becomes tender, smear it with your palm and apply the liquid on the chest, neck, around the nostril and finally leaving the moist leaves above the head. Repeat this process 3-4 times in a day for 2-3 days.
78	*Lantana camara* Linn. [Verbenaceae] Big/wild sage	*Aring toh/ Thirak Lei* [Sh]	*Nongban Lei* [P]	Leaf [Raw] – [E]	Blood Clotting/Cuts and Wounds, Freckles, Acts as Anti-fungal, Microbial and Insecticidal: When a person suffers from minor cuts and wounds crushed some fresh leaves and immediately apply the paste on the cuts area. This will stop from bleeding as it helps in blood clotting and heal fast. The leave of this shrub plant is also believed to act as anti-fungal, microbial and insecticidal. Crushed some fresh leaves and rub with the paste on the freckles or red pigmentation on the face and let it is dry for awhile. Apply twice a day for a week.
79	*Leucas aspera* (Willd.) Linn. [Lamiaceae] Thumbai	*Wanchei ann* [H]	*Mayang yenbum* [S]	Whole plant [O]	Blood enhancement (Post-natal woman), Sinus, Constipation: Boil some fresh tender leaves as simple cooked curry along with fresh fishes and have regularly for 5- 7 days or occasionally to enhance blood especially for post-natal woman. It also suggested it helps cured sinusitis and constipation persons.
80	*Lindernia ciliata* (Colsm.) Pennell [Linderniaceae] Hairy Slitwort	*Kihom maan* [H]	*Kihom maan* [S]	Whole plant [O]	Stone cases (Kidney/Gall bladder): Some amount of this *Lindernia ciliata* creeper plant is boil in about 2/3 litres of water and taken a glassful 3 times a day for 5-7 days or till cured.

Sl. No.	Botanical Name [Family]- Common Name	Chothe Name- [H, Sh, T, C]	Manipuri Name [S/P]	Parts Used [Forms]- [O/E]	Diseases – Treatment Methods and its Dosages
81	*Litsea cubeba* (Lour) Pers. [Lauraceae] Mountain pepper, Litsea	*Singngathai* [S]	*Usingsha mapan* [S]	Fruit, Flower [O]	Asthma, Cool. Rheumatism: Some amount of the fresh fruit is crushed and eaten to eliminate asthma, cool and rheumatism once or twice a day or occasionally to dispel the unwanted oxidant elements in the body due to the pepper like components. About 2 ml of the extracted oil of the fruits is taken once or twice a day for a week or occasionally when required by those persons suffering from asthma, cool and rheumatism.
82	*Litsea glutinosa sebifera* Thumb. [Lauraceae] Brown/Soft Bollywood	*Thaang hidak* [T]	*Thaang hidak panbi* [P]	Bark, Root [E]	Fracture and Sprain: Crush the bark properly and bandage the gummy paste by adding little water around the fracture or sprain part for 2-3 days. Repeat the process by removing the old paste and applying afresh again after 2 days.
83	*Ludwigia adscendens* (L.) H.Hara [Onagraceae] Water primrose (white)	*Tui kundo* [C]	*Esing kundo* [P]	Leaf [O]	Diabetes: Cooked the fresh leaves along with fresh local dog fish (*Ngakara*) as simple curry and serve it twice a day (Morning and evening) for atleast for 3-5 days and may continue occasionally if helpful till cured.
84	*Lygodium japonicum* (Thunb.) Sw. [Schizaeaceae] Japanese climbing fern	*Lai chang-khrang* [C]	*Lai chang-khrang* [P]	Whole plant [O]	Jaundice: Collect the ashes of the burn dried creeper plants and filter the ash water in a funnel and add little *sitamai* (crystal cane sugar) and take the cold decoction solution half a glass twice a day for 7 days or more after food till cured.
85	*Lysinachia candida* Lindl. Primulaceae] Manipur loosestrife	*Kengoi* [H]	*Kengoi* [S]	Whole plant [O]	Anaemia: Boiled the whole plant of tender stems along with leaves and served as decoction half a glass twice a day for a week. One may also cook as simple curry along with dried fishes and served occasionally.

Sl. No.	*Botanical Name [Family]- Common Name*	*Chothe Name- [H, Sh, T, C]*	*Manipuri Name [S/P]*	*Parts Used [Forms]- [O/E]*	*Diseases – Treatment Methods and its Dosages*
86	*Melastoma malabathricum* Linn. [Melastomataceae] Matakui (Palau)	*Joubung* [H]	*Yachubi* [S]	Leaf [O]	(i) Jaundice, (ii) Diabetes: i) Boiled the fresh leaves and take half a glass of the decoction twice a day for 7 days after food only. It will be milky and sticky after boiling it. ii) Boil some fresh leaves and take a glassful of the decoction 3 times a day for a week or continue till the diabetes millipedes (frequent urine excretion) is reduce. It is suggested to have the decoction solution instead of water whenever one feels thirsty. This is for those diabetes persons suffering from frequent urine release.
87	*Melothria perpusilla* (Blume.) Cogn. [Cucurbitaceae] Pubasari, Wild cucumber	*Lam sakma* [C]	*Lamthabi* [S]	Whole plant/ Tuber [O]	Jaundice, Mal-nutrition: Boil some of the whole plant either fresh or dried in about 2 litres of water and take a glassful of the decoction 3 times a day for 5-7 days to cure from jaundice. One may also take the decoction occasionally. Even the fruit are used in the preparation of the medicine. One can use the tuber when not available the leaf.
88	*Menthe arvensis* Linn. [Lamiaceae] Spearmint/Pudina	*Nungsi hidak* [H]	*Nungsi hidak* [P]	Leaf [Raw]- [O]	Stomach ache, Gastric, Antipyretic: Person suffering from stomach ache, gastric and antipyretic cases may take about 10-20ml of the extracted liquid of the crushed fresh leaves (or may add little honey) twice a day till cured.
89	*Meyna laxiflora* Robyns. [Rubiaceae] Muyna/Helu/Alu	*Theipi* [T]	*Heibi* [S]	Leaf and dry fruit [Raw]- [O]	i) Blood purification, ii) Constipation: i) To enhance blood purification of a person the fresh leaves may be eaten as chutney time to time, considering rich in iron contents. ii) One may eat plenty of the dried ripe fruits if one is having constipation problem.

Sl. No.	*Botanical Name [Family]- Common Name*	*Chothe Name- [H, Sh, T, C]*	*Manipuri Name [S/P]*	*Parts Used [Forms]- [O/E]*	*Diseases – Treatment Methods and its Dosages*
90	*Millettia pachycarpa* Bentham. [Fabaceae] Fish poisoning	*Ngamu jai* [Cl]	*Ngamu jai* [P]	Stem, root [E]	(i) Dog flea, (ii) Fish poison: i) Cut a piece of the mature stem or root and split into two halves and wear it around the dog's neck. The dog's flea will get rid itself of the smell. ii) Crush some roots and put it in the river cleavages or stream, after half an hour or an hour the fishes will appear on the surface giddy struggling to survive.
91	*Mimosa pudica* Linn. [Fabaceae] Sensitive plant/ (Touch-me-not)	*Ajakpi thingna* [H]	*Kangphai ikaithabi* [P]	Leaf, root [Raw] – [O]	Post-delivery Uterine pain: Woman suffering from post-delivery uterine pain may take about 10-20ml of the liquid obtained by crushing the leaves and roots as tonic (or mixed with honey) before meal twice a day till one gets relief. Sometimes the whole plant is boiled and served as decoction a glassful before food twice a day for a week or more till cured.
92	*Mussaenda roxburghii* Hook. F. [Rubiaceae] East Himalayan Mussaenda.	*Sunubarei* [Sh]	*Hanurei* [S]	Leaf, Root [E]	Boil, Cancer, Tumour: Boil some fresh leaves and bandage around the boil or infected tumour areas for 3-5 days by dressing the wound every day.
93	*Musa sapientum* Linn. [Musaceae] Banana stem/ Pseudostem and flower	*Changlong* [Sh]	*Laphu* [P]	Stem and Flower [Raw, as concoction] – [O]	i) Chronic ulcer, ii) Lactation of mother, iii) Constipation: i) Person who suffers from prolonged chronic ulcer may drink about half a glass (60ml) of the extracted liquid of the edible young fresh banana stem (*Pseudo stem*) in empty stomach once a day for a week or more. ii) Young mother who lack breast milk may cooked along with some dried fishes and eat regularly for a week or frequently time to time to improve the breast milk for the baby. Therefore, this young banana stem especially grown from the hills are most preferred by the locals. iii) Some recommended that eating the simple cooked banana stem with dry fishes or as chutney helps reduced constipation if taken occasionally.

Sl. No.	*Botanical Name [Family]- Common Name*	*Chothe Name- [H, Sh, T, C]*	*Manipuri Name [S/P]*	*Parts Used [Forms]- [O/E]*	*Diseases – Treatment Methods and its Dosages*
94	*Nicotiana tabacum* Linn. [Solanaceae] Tobacco plant	*Thakna* [Sh]	*Hidak Mana* [S]	Leaf [Raw or dried] – (E)	Acts as insecticides, Itchy, allergies, insects bite, leech bite: Traditionally, when a person is bitten by an insect or leach, feels itchy, experience allergy one usually smear the tobacco leaf with water or spit and is apply on the infected area. The application of tobacco leaf on infected area acts as anti-allergy and repellant.
95	*Nyctanthes arbor-tristis* Linn. [Oleaceae] Night flowering Jasmine, Parijat	*Singgharei* [Sh]	*Singgharei* [S]	Leaf [O]	Malaria fever: Crushed the fresh leaves and add little honey in the extracted liquid and take about 10-20ml twice a day for 2-3 days before food. (For children give less dose about 1-5ml). Avoid any fresh red meat during medication.
96	*Ocimum sanctum* Linn. [Lamiaceae] Holy basil/Krishna tulsi	*Tulsi* [H]	*Tulsi* [S]	Leaf [Raw] – [O]	Cough, Bronchitis, Tonsillitis, Asthma: Person suffering from most air-borne diseases like cough, bronchitis, tonsillitis, asthma and congestion of lungs may take about 10-15ml of the extracted leaves juice mixed with little honey 3 times a day till it is cured. For children below the age of 3 years suffering from cough, add some fresh leaves in some amount of boiling water and give 2-3 tea spoon of the decoction 3 times a day till relief.
97	*Ocimum tenuiflorum* Linn. [Lamiaceae] Thai holy basil, Tulasi	*Tulsi ahangpa* [H]	*Tulsi amuba* [S]	Leaf [E]	Sinusitis, Cough: Crushed or smeared some fresh leaves with your palm and keep inhaling the smell time to time regularly 3-4 times a day for 7 days or till cured.
98	*Oenanthe javanica* Blume. DC. [Apiaceae] Java water dropwort/ Chinese celery/Water celery	*Komprek* [H]	*Kongprek* [S]	Leaf [Raw] – [O]	Hypertension, Indigestion, Blood Pressure: Person suffering from hypertension, high blood pressure and indigestion may take about 20-30ml of the extracted liquid of the crushed fresh leaves of *Oenanthe javanica* by adding little water (and honey) or half a glass after mixing with little water before food or after as tonic twice a day till cured. However, many preferred the decoction style of medication *i.e.* by boiling certain amount of the leaves and stem and serving a glassful twice a day before or after food and also eating the boiled leaves and stems along with other food items.

Sl. No.	*Botanical Name [Family]- Common Name*	*Chothe Name- [H, Sh, T, C]*	*Manipuri Name [S/P]*	*Parts Used [Forms]- [O/E]*	*Diseases – Treatment Methods and its Dosages*
99	*Oroxylum indicum* Linn. Kurz. [Bignoniaceae] Indian trumpet flower	*Maklong* [T]	*Shamba* [S]	Bark, fruit and leaf [Raw] – [O/E]	i) Piles, ii) Muscles/Joint pains, high Blood pressure and, iii) Epilepsy: i) Person suffering from piles may sit on the warm boiled water of the plant bark kept in a tub. This process is continued till one gets relief. The plant bark should be peeled upwards with one breath. For piles even the young tender fruit is slightly roasted and taken twice a day till cured. ii) Person who suffers from muscles/joint pains may take a bath with the boiled water of the bark or leaves. Or take a small towel and dipped into the boiled hot water of the leaves/bark, and balms the joint pain areas till relief. iii) Those suffering from Epilepsy may either eat the tender leaves cooked as curry or the fruit slightly roasted occasionally till cured.
100	*Oxalis corniculata* Linn. [Oxalidaceae] Creeping woodsorrel/ procumbent yellow sorrel	*Yensin arikpi* [H]	*Yensil* [P]	Leaf/Whole plant [Raw] – [O]	i) Indigestion, Gastric, ii) Blood purification: i) Person suffering from indigestion or gastric may take about 20-30ml of the extracted liquid of the whole plant before food by adding little water and honey or salt twice a day for a week or till it is cured. ii) Boiled the whole plant in about a litre of water and one may add little salt and take the decoction half a glass or full glass twice a day for a week or time to time to enhance blood purification. It is suggested that during medication one should avoid fatty or oilly things, chilly or spicy items.
101	*Paederia foetida* Linn. [Rubiaceae] Stinkvine.	*Yoisukjam* [C]	*Oinum* [S]	Leaf [E]	Fracture (esp, of bird and animals): Crush the fresh leaves and apply the paste of *Paederia foetida* especially on animals and birds wounded area.

Sl. No.	*Botanical Name [Family]- Common Name*	*Chothe Name- [H, Sh, T, C]*	*Manipuri Name [S/P]*	*Parts Used [Forms]- [O/E]*	*Diseases – Treatment Methods and its Dosages*
102	*Parkia timoriana* (DC.) Merr. [Fabaceae] Tree bean (Syn. *Parkia timoriana*)	*Jongtak* [T]	*Yongchak* [S]	Seed [Dried] – [O],	i) Chronic ulcer, Dysentery, ii) Diarrhoea: i) Person suffering from prolong severe chronic-ulcer boiled some amount of the dried outer covering of the long tree bean fruits (*Parkia roxburghii)* in about a litre of water for half an hour or more till the water becomes deep red. When it is cold take half a glass of the decoction twice before food regularly for a week or more till it reduces. The bitter and clumsy taste help relief the pain and thickens the stomach and intestinal linings. ii) For person suffering from dysentery and diarrhea may also take half a glass of the decoction twice a day till relief of the problem. During medication one is suggested to avoid hot, spicy and oily food items.
103	*Phaseolus lunatus* Linn. [Fabaceae] Butter bean, Lima bean	*Kalandri* [Cl]	*Kalandri* [S]	Leaf [E]	Fever, Hot head: Crush the fresh leaves and apply the paste above the head and around the pituitary glands areas time to time till cured.
104	*Phyllanthus urinaria* Linn. [Phyllanthaceae] Chamber bitter, Gripeweed, Stonebreaker	*Lam theichura arapa* [H]	*Chakpa heigru* [S]	Whole plant [O]	Diabetes: Boil certain amount of the fresh whole plant and serve the decoction solution a glassful three times a day for a week. If there is a sign of improvement, prepare the decoction and take according to once requirement once a day or twice a day. Actually this solution is a replacement of water. The diabetes patient should drink the solution instead of water whenever thirsty.
105	*Phyllanthus emblica* Linn. [Phyllanthaceae] Indian Gooseberry, Amla	*Theichuraa/ Shuru* [Sh]	*Heikru* [S]	Bark and fruit [Fresh/Dry]- [O/E]	i) Cough and Cold, ii) Sore Eye, iii) Body stimulant: i) Person suffering from cough and cold, and those feeling weaknesses may crushed the fresh goose berries and take about 10-20ml 3-4 times in a day till relief. For cough some prefer to roast abit the goose berry in the fire have as much as one can time to time till cured. ii) For sore eyes few drops of the extracted juice of the gooseberry may be poured into the sore eye till one gets cured.

Sl. No.	Botanical Name [Family]- Common Name	Chothe Name- [H, Sh, T, C]	Manipuri Name [S/P]	Parts Used [Forms]- [O/E]	Diseases – Treatment Methods and its Dosages
					iii) The extracted liquid obtained from the crushed fresh bark is mixed with little water and taken as body stimulant substance. The goose berry is widely grown in the deep forest while some planted around their courtyard.
106	*Plantago major* Linn. [Plantaginaceae] Broad plantain, white man's foot, Greater plantain	*Antapot* [H]	*Yampat* [S]	Leaf/stem [Raw] [E]	i) Boils, ii) Wounds: i) Person suffering from boils may heat up little the fresh leaves on the fire till it gets decoloured. Then, roll abit and applied on the boil part with little opening on the tip. Repeat this process till the boil is matured and the pus comes out or healed. ii) Crushed the fresh leaves and apply the paste on the wound. Repeat it time to time.
107	*Polygonum odoratum* Lour. [Polygonaceae] Smartweed/Laksa plant/Vietnamese mint	*Phakphai* [H]	*Phakpai* [S]	Leaf [Raw or dried] – [O]	Hypertension, Tonsils and Throat problems: Person suffering from hypertension, tonsillitis and other throat problems may take about 20-30ml of the fresh extract liquid of the leaf mixed with little honey as tonic for 3/4 days before food.
108	*Pouzolzia indica* (L.) G. Benn. [Urticaceae] Bo mam, Apoyapoyan, Graceful Pouzolzsbush	*Muithemte* [H]	- [S]	Tuber [E]	Boil, Cancer (Tumour on thigh, arm and muscle areas): Crushed the rhizome or tuber of this plant and apply the gummy paste around the boils. For cancer treatment it should be used only in the advance stage of the cancer tumour part) and bandage it for 2 days and repeat it again the process with fresh tuber till cured. The utensil used should be sterilised before used. (It is strongly suggested to avoid using any sharp metallic objects on the infection or while dressing it).
109	*Prunus persica.* Linn. [Rosaceae] Peach	*Chumprei* [T]	*Chumbrei* [S]	Leaf [E]	Wounds and Worms, Anti-wormicide: If a wound is infested with worms and maggots or became septic the fresh matured leaves of *Prunus persica* are crushed and the paste is applied on the wound (and the liquid may be squeezed into the wound before applying). This treatment is usually used on animals like pig, cow and buffalo when infected with maggots.

Sl. No.	*Botanical Name [Family]- Common Name*	*Chothe Name- [H, Sh, T, C]*	*Manipuri Name [S/P]*	*Parts Used [Forms]- [O/E]*	*Diseases – Treatment Methods and its Dosages*
110	*Psidium guajava* Linn. [Myrtaceae] Guava	*Bareitun* [T]	*Pongtol* [S]	Leaf and fruit [unripe fruit] – [O]	Dysentery and Diarrhoea: Traditionally, person suffering from dysentery or diarrhea may eat some fresh tender leaves of guava (with little salt) time to time till relief. Though little bitter and clumsy in taste it greatly helps in suspension. Eat plenty of unripe guava fruit if one is suffering from dysentery and diarrhea (with little common salt if needed be). However, excessive eating of upripe guava leads to constipation.
111	*Punica granalum* Linn. [Onagraceae] Pomegranate	*Kapo thei/ Theisan* [Sh]	*Kaphoi/ Kapohei* [S]	Leaf, fruit [O]	Dysentery/Diarrhoea: Chew the fresh tender leaves of the local pomegranate as much as possible to help suspend the diarrhoea or dysentery. One is suggested to keep chewing the fresh leaves whenever one feels like any time of the day.
112	*Rhus succedanea* Linn. [Anacardiaceae] Wax tree/ Wild varnish dry galls	*Hokma* [Sh]	*Heimang* [S]	Leaf and Fruits [Raw and dried] – [O]	i) Dysentery, Indigestion, ii) Constipation, iii) Body stimulant: i) Person suffering from dysentery and indigestion may eat some of the fresh tender leaves of *Rhus succedanea* (mixed with little salt) till it is suspended. Or about half/full glass of the boiled dry fruit of the decoction (mixed with little salt and sugar), when completely cold is taken 2/3 times a day till one gets relief. ii) On the other hand, those persons suffering from constipation may drink the grind seeds powder mixed in hot water along with little chilly and salt, that served as enema. iii) When one feels lethargic or dull or lacks body stimulation add a tea spoon or two of the grind dry seeds powder in a litre of boiling water and served half/full glass of it. Or one may add a teaspoon or two while preparing red tea and also add sugar to one's taste and have either regularly or occasionally when needed to energise or enhance body stimulation especially after heavy meal.

Sl. No.	*Botanical Name [Family]- Common Name*	*Chothe Name- [H, Sh, T, C]*	*Manipuri Name [S/P]*	*Parts Used [Forms]- [O/E]*	*Diseases – Treatment Methods and its Dosages*
113	*Ricinus communis* Linn. [Euphorbiaceae] Castor bean, Castor oil plant	*Keke ashangpa* [Sh]	*Keke anganba* [P]	Leaf [E]	Giddiness (esp. woman): Boil some fresh leaves along with the local shampoo (*Chinghe*) and wash the head and hair regularly for a week or occasionally with the solution to get relief from the giddiness or morning sickness type.
114	*Rorippa Palustris* (L.) Besser. [Brassicaceae], Indian yellowcress	*Aju antam* [H]	*Uchi hangam* [S]	Leaf, whole plant [O]	Diabetes, Dysentery: The Fresh plant is cooked and eaten regularly against diabetes. The whole plant is boil and a glassful is served twice a day for two days those suffering from dysentery.
115	*Sacharum officinarum* [Poaceae] Sugar cane	*Maanshu/ Molshu* [Sh]	*Chu* [S]	Stem [Fresh juice] – [O]	Jaundice: Person suffering from jaundice may drink 1 or 2 glassful of freshly extracted sugar cane juice 3-4 times daily, especially in the morning for 2-3 weeks till one gets cured from the jaundice.
116	*Sapindus trifoliatus* Linn. [Sapindaceae] South India Soapnut, Soap berry	*Lingsi* [T]	*Kekru* [S]	Fruit [E]	Fever (esp. children): Soak half of the dried seed for a while in a half a mug of water and stir it with a spoon till it gives a thick foam then apply the foam above the head time to time till the fever subside and reduced the body temperature.
117	*Schefflera arboricola* (Hayata) Kanehira [Araliaceae] Dwarf umbrella plant	*Tarang keikhomei* [Cl]	*Utangbi* [P]	Leaf [E]	Body/Muscle/Joint pains: Boil the fresh leaves of *Utangbi* (nine leave variety), along with *Nungshangpanbi* and *Nongmangkha* and used the warm decoction solution as balm by giving a sponge body massage with it twice a day for 3 days or occasionally to relief from muscles and body ache.
118	*Scutellaria discolor* Colebr. [Lamiaceae] Skullcap, Daamapaate/Nilo	*Ann lakhat* [H]	*Yenakha/Ching yenakha* [S]	Whole plant [E/O]	(i) Irregular menstrual problems (removal of bad blood), (ii) Malaria, (iii) High blood pressure, (iv) Fever (esp. children): i) Crush some fresh leaves and stems (whole plant) along with a piece of dry tobacco leaf and shock it with little pure alcohol (liquor) for two days, (plus few drops of tobacco liquid) after which, apply the concentrated liquid upwards from the feet and downwards from the head to shoulders, chest and back (but not around the stomach or belly area).

Sl. No.	Botanical Name [Family]- Common Name	Chothe Name- [H, Sh, T, C]	Manipuri Name [S/P]	Parts Used [Forms]- [O/E]	Diseases – Treatment Methods and its Dosages
					ii) Boil the whole plant and add little honey and have about 5-10ml of the decoction solution 2 times a day for 3-5 days or till cured. (Avoid take more than the required amount). For children give about 1-2ml only. iii) Boil the fresh leaves along with black sticky rice and tobacco extracted liquid (Thakpu tui) and use the solution as balm from head to shoulder and feet to hip twice a day for 3 days or till reduced. iv) Crush the fresh leaves on your palm and add little saliva and apply the paste on the child's body, chest and back to reduce the fever.
119	*Sida rhombifolia* Linn. [Malvaceae], Rhombus-leaved sida, Paddy's lucerne.	*Think ulin* [H]	*Uhan* [S]	Leaf [E]	Worms (esp. for children): Crush the leaves and apply the paste on the baby's or child head. When the paste becomes dry due to body heat then, add few drops of water on the paste to just wet it. If not so effective repeat the process once again but not too frequently.
120	*Solanum indicum* Linn [Solanaceae] Yellow berried nightshade (small variety)	*Samtok/ Samtrok arikpa* [Sh]	*Leipungkhang* [S]	Fruit [Raw,] – [O]	i) Piles, ii) Measles, iii) Cough: i) The raw fruit is crushed and mixed with little honey and taken before each meal. Person with serious pile case that bleeds may take three times in a day, till it cures. ii) Babies and children suffering from measles may take the crush fresh fruits mixed with little honey time to time till it completely expels out the virus and is cured. iii) The young fruits are steamed and prepare as chutney or added in simple local dishes to prevent from cough and other ailments.
121	*Solanum virginianum* Linn. [Solanaceae] Kantakari, Surrattense nightshade, Yellow-fruit nightshade,	*Pheisouthai* [H]	*Japan kamen* [P]	Seed, Fruit [O]	(i) Tooth ache, (ii) Animals/Cow leach (*Lingaret*): i) The seeds of the ripe fruits are extracted and kept it dry. Then burn the dried black seeds with the matches in a clean container and inhale the smoke through the mouth focusing towards the toothache. It is believed that the germs of the toothache fall down after a while. Repeat this burning the dried seeds and inhaling the smoke 3-4 times till the tooth ache is relief.

Sl. No.	Botanical Name [Family]- Common Name	Chothe Name- [H, Sh, T, C]	Manipuri Name [S/P]	Parts Used [Forms]- [O/E]	Diseases – Treatment Methods and its Dosages
					ii) Crush some of the ripe fruits well and pour it into the animal's nostril like cow or buffalo infested with parasitic worm or leach. Within few minutes the infested leaches will exit itself from the animal's nostril.
122	*Tamarindus indicus* (L.) [Caesalpiniaceae] Tamarind	*Mangke* [T]	*Mangge* [S]	Fruit [O]	Measles: Apply just 2/3 drops of the ripe fruit of tamarind jam on the tongue and lips of the measles infected child. And later if the child demands for any craving items, provide and feed him to his satisfaction.
123	*Thalictrum foliosum* DC. [Ranunculaceae] Gold thread root, Pilijari	*Changkrang aroupa* [C]	- [P]	Leaf [E]	Sprain: Crush some amount of the fresh leaves into paste and apply it around the sprain or fracture part for 2 days and repeat the process till it is cured.
124	*Toona ciliate* M. Roem. [Meliaceae] Red cedar	*Atithing/Tairen* [T]	*Tairen* [P]	Leaf [O]	(i) Typhoid, (ii) Cough, (iii) Malaria fever (iv) Measles and Small pox, (v) Allergy: i) Crush some leaves of *Tairen mana* and add some honey and have a teaspoonful twice a day after food till the typhoid is reduced. (Pluck the leaf that has single leaf or three leaves at the tip of the stalk). ii) Boil some amount of the fresh leaves and served about 10-20ml of the decoction early morning in empty stomach, have 3 times a day for 3-5 days. One may also boil separately certain amount of the fresh leaves and take a bath with the decoction, if one does not have fever. iii) Boil some fresh leaves and take about 10-20ml of the decoction solution 3 times a day in empty stomach for 3-5 days in case of person suffering from malaria fever. (Do not take bath during fever). iv) Boil some fresh leaves and take about 10-20 ml in case of measles or small pox of the decoction solution 2 times a day for 3 days or more. (For children the dosage should be 1-5ml). After fully cured, one may take bath with the decoction solution boiled separately. (v) Collect some fresh leaves and dry fry it till it turns harden and blacken, then crush into powder form and add little pure mustard oil and apply the paste over the infected body parts twice a day for 2-3 days.

Sl. No.	*Botanical Name [Family]- Common Name*	*Chothe Name- [H, Sh, T, C]*	*Manipuri Name [S/P]*	*Parts Used [Forms]- [O/E]*	*Diseases – Treatment Methods and its Dosages*
125	*Trachyspermum roxburghianum* DC. Craib. [Apiaceae] Wild celery/Ajmod	*Ching Pathikhom* [H]	*Ching pathikhom* [S]	Leaf/whole plant [O]	Enhance Blood Purification: Person who wants to enhance blood purification may take about 10-20ml of the extracted liquid obtained by crushing the fresh leaves mixed with little honey twice a day before each meal for a week or more till satisfied.
126	*Vitex negundo* Linn. [Verbenaceae] Chaste tree	*Jenlik shibi* [Sh]	*Urik shibi* [S]	Leaf [E]	i) Allergy, Scabies, ii) Chicken fleas: i) Boil some fresh leaves and take a bath with the warm solution for 3-5 days to cure from allergy and scabies. ii) Stick some branches inside and around the poultry to dispel the chicken fleas. It acts as insecticides.
127	*Viola pilosa* Blume. [Violaceae] Viola, Banafshah	*Shantang leipak* [H]	*Huikhong* [S]	Leaf [O]	To enhance blood and stimulates body. The fresh leaves are either boiled or eaten afresh along with local chutney occasionally to enhance blood and stimulate the body.
128	*Zanthoxylum acanthopodium* [Rutaceae] Pricky winged leaf/ Darmar/Sheguan pepper	*Singree* [Sh]	*Mukthrubi* [S]	Leaf and seed [Fresh leaf and dry seeds] – [O/E]	Cough, Asthma and Arthritis, Food poison: Person suffering from cough, arthritis and asthma may boiled the leaves and the steam and inhaled the steam and one may also take about 20ml of the decoction twice a day for a week or occasionally till relief. When leave is not available, one may boil some amount of the brown seed covering in half a litre of water and served about 20-30ml when warm to relief from dry cough till relief. (It is suggested to avoid sweats and oily items during medication). Crushed some fresh leaves and eat as much as possible to relief from food poison of lethargy.

Sl. No.	Botanical Name [Family]- Common Name	Chothe Name- [H, Sh, T, C]	Manipuri Name [S/P]	Parts Used [Forms]- [O/E]	Diseases – Treatment Methods and its Dosages
129	*Zea mays* L. [Poaceae] Maize	*Chungachak* [Sh]	*Chujak* [S]	Fruit [O]	(i) Food poison, (ii) Wrong medication: i) Crush the young maize into powder form (by adding little water) and immediately give the liquid paste form as much as possible to the patient. Give the solution several times even after the person vomited many times till it is fully cleanse and neutralised. (In certain case, when the fresh maize is not available, get the dry bud of the maize and crush it into powder form by adding little water and force the person to drink as much as possible till neutralised). ii) The same solution may be given to person in certain cases when a person took pesticides, insecticides, or wrong medicine or prescription.
130	*Zingiber officinale* Rosc. [Zingiberaceae] Ginger	*Aithing* [H]	*Shing* [S]	Rhizome [Raw] – [O/E]	i) Cough, Bronchitis, ii) Piles, iii) Divination Purposes: i) When a person suffers from cough and bronchitis, one may chew pieces of fresh ginger time to time till one gets cured. About 5-10ml of the extracted liquid mixed with little salt may be taken before food till one gets cured. ii) Sometimes to enhance body stimulation, some pieces of fresh ginger may be added while boiling red tea and taken along with tea to enhance body stimulation or increase immune system. iii) When an adult suffers from constipation, he may insert a small amount of crushed fresh ginger mixed with little mustard oil or burnt kerosene residue inside his anus to remove the gas and old stole. Some believed that it even helps from the minor piles problems if done occasionally.

Abbreviations: Linnaeus (Linn.), Herbs (H), Shrub (Sh), Tree (T) Climber (C); Seasonal (S), Perennial (P).

Table 2: Frequency of the Plant's Family

Sl. No	*Plant Family*	*Frequency*	*Sl.No*	*Plant Family*	*Frequency*
1	Acanthaceae	2	30	Linderniaceae	1
2	Acoraceae	1	31	Malvaceae	4
3	Alliaceae	1	32	Meliaceae	3
4	Amaranthaceae	3	33	Melastomataceae	1
5	Anacardiaceae	1	34	Moraceae	3
6	Annonaceae	1	35	Musaceae	1
7	Apiaceae	4	36	Myrtaceae	1
8	Araceae	2	37	Oleaceae	1
9	Araliaceae	3	38	Onagraceae	2
10	Asphodelaceae	1	39	Oxalidaceae	2
11	Asteraceae	7	40	Phyllanthaceae	4
12	Balsaminaceae	1	41	Plantaginaceae	1
13	Basellaceae	1	42	Poaceae	3
14	Bignoniaceae	1	43	Polygonaceae	1
15	Brassicaceae	1	44	Primulaceae	1
16	Caesalpiniaceae	2	45	Ranunculaceae	1
17	Cannabaceae	1	46	Rosaceae	1
18	Capparidaceae	1	47	Rubiaceae	4
19	Caricaceae	1	48	Rutaceae	3
20	Caryophyllaceae	1	49	Sapindaceae	1
21	Convolvulaceae	3	50	Saururaceae	1
22	Cornaceae	1	51	Schizaraceae	1
23	Costaceae	1	52	Solanaceae	4
24	Cucurbitaceae	2	53	Ulmaceae	1
25	Dipterocarpaceae	1	54	**Urticaceae**	2
26	Euphorbiaceae	3	55	**Verbenaceae**	2
27	Fabaceae	8	56	Violaceae	1
28	Lamiaceae	11	57	Vitaceae	1
29	Lauraceae	3	58	Zingiberaceae	9
				Total	**130**

Table 3: Different Ailments and Diseases

Sl. No.	*Ailments and Diseases*	*Freq*	*Sl.No*	*Ailments and Diseases*	*Freq*
1	White discharge of woman	3	36	Gonorrhoea	1
2	Urinary problems	3	37	Liver and Hepatic problems	2
3	Diarrhoea (Esp. for children)	4	38	Tonsillitis, Bronchitis	4

Sl. No.	*Ailments and Diseases*	*Freq*	*Sl.No*	*Ailments and Diseases*	*Freq*
4	Worms/Ascaris (esp. for children)/Intestinal worms	5	39	Virus infections/Skinny, Pricked by poisonous plants like boils	1
5	Jaundice	7	40	Cuts and wounds	7
6	Fever	9	41	Giddiness/Migraines (esp. woman)	4
	Burn	1	42	Irregular menstrual problems	6
7	Paralysis	1	43	Rheumatism:	2
8	Enhance blood (for Post-natal woman)	2	44	Fracture and Sprain	4
9	Asthma	5	45	Toothache	3
10	Termination of Conception of a child (Abortion)	2	46	Dark facial pigmentation	1
11	Extra tissue growth, Haemorrhoid		47	Chronic-ulcer	6
12	Piles	6	48	Anti-fungal, Microbial and Insecticidal	1
13	Indigestion	4	49	Germicides	1
14	Urinary problems		50	Reduce weight	1
15	Diabetes	8	51	Stomach ache	4
16	Mud stain on nails	1	52	Dog flea	1
17	Cough, Cold	11	53	Fish poison	1
18	Boil	9	54	Post-delivery Uterine pain	1
19	Anaemia, Blood purification	9	55	Cancer, Tumour	3
20	Giddiness, Nausea	4	56	Lactation of mother	1
21	High Blood pressure	6	57	Typhoid	2
22	Kidney failure, Urinary infectious problems, Stone kidney problems	9	58	Malaria fever	3
23	Dysentery	6	59	Sinusitis	2
24	Diarrhea	4	60	Indigestion	4
25	Gastritis, Gas formation	7	61	Epilepsy	1
26	Toothache	3	62	Measles	4
27	Stone cases (Kidney) :-		63	Allergy	4
28	Constipation	5	64	Crack on the lips and tongue	2
29	Hypertension	5	65	Nervous spasms	1
30	Dispeller of Magical Spell/To ward off evil spirits	3	66	Complexion	1
31	Ringworm and Scabies	2	67	Scare marked, Bruise	1
32	Conjunctivitis, Sore eyes, Ear liquid	3	68	Arthritis	1
33	Joint pains (Muscular and bones)	4	69	Food poison, Wrong medication	1
34	Smallpox	3			
35	Carminative Stimulant	1			

Table 4: Forms of the Plants

Sl. No.	*Forms of Plant*	*Frequency*
1	Tree	22
2	Shrub	34
3	Herb	59
4	Creeper	9
5	Climber	6
	Total	130

Table 5: Types of the Plants

Sl. No.	*Types of Plant*	*Frequency*
1	Seasonal	99
2	Perennial/Annual	31
	Total	130

Table 6: Parts of the Plant Used

Sl. No.	*Parts of Plant Used*	*Frequency*
1	Leaf	84
2	Whole plant	15
3	Fruit	19
4	Seed	5
5	Bark	6
6	Stem	13
7	Root	6
8	Rhizome	8
9	Flower	6
10	Bulb	1
11	Tuber	1
	Total	164

Table 7: Treatment Methods

Sl. No.	*Treatment Method*	*Frequency*
1	Oral	90
2	External	49
3	Dual/Both	09
	Total	148

Table 8 Ethnomedicinal Plants that are Eaten as Food Habits

Sl. No	Botanical name [Family]- Common Name	Chothe Name-	Manipuri Name	Parts used [Forms]- [O/E]
1	*Acemella oleracea* (L.) R.K. Jansen. Toothache plant	*Ansa*	*Morok sabi*	Leaf – Plain boiled or cooked as simple curry with dried fishes.
2	*Adhatoda vasica* Nees. Malabar nut	*Annha angoupa*	*Nongmangkha Angouba.*	Leaf – Plain boiled.
3	*Aleo vera* (L.) Burm. f. Aleo vera	*Kheva arikpa*	*Aleo vera*	Leaf – Gel eaten raw.
4	*Allium sativum* Linn.	*Satun*	*Chanam*	Bulb – Added in curry.
5	*Alocasia indica* (Roxb.) Schott. Giant Taro	*Jaitim*	*Yendem*	Stem – Added in curry or as chutney (*Eromba*).
6	*Alpinia caerulea* (R.Br.) Benth. Native ginger	*Phuiphung*	*Pulei manbi*	Root – Added in fish and beef curry.
7	*Alpinia galangal* Linn. Willd. Greater galangal	*Marou (Brou)*	*Kanghu*	Rhizome – Dried powder eaten as chutney.
8	*Amaranthus retroflexus* Linn. Spiny amaranth/Spiny pigweed	*Si-ake anbu Arintoipa*	*Chengkruk (Tingkhangpanbi)*	Leaf – Cooked as simple curry with dried fishes.
9	*Antidesma acidum* Retz. Amita	*Tuitrit*	*Ching Yensin*	Leaf - Cooked as simple curry with dried fishes or meat.
10	*Averrhoa carambola* Linn. Star fruit	*Heinouchom*	*Heinoujom*	Fruit – Cooked as pickle with sugar.
11	*Bauhinia variegate* (L.) Benth. Orchid tree, Mountain-ebony	*Chingthrao/ Heina kundo*	*Chingthrao/Heina kundo*	Leaf – Cooked as simple curry with fishes.
12	*Benincasa hispida* Thunb. Cogn. Ash-gourd/Winter melon	*Maipoy/ Mayngou*	*Torbot*	Fruit – Plain boiled or cooked with duck meat and pork.
13	*Bischofia javanica* Blume. Bishop wood	*Uthum toh*	*Uthum panbi*	Leaf (Tender) – Plain boiled and taken along with chutney.
14	*Brassaiopsis hainla* (Buch.-Ham.) S Chuletro, Seto Chuletro Tree	*Chom*	*Chom (variety)*	Fruit – Steam and eaten along with chutney.
15	*Cajanus cajan* (Linn.) Millsp. [Fabaceae], Pigeon pea	*Meirongpi*	*Meirongbi*	Leaf – Plain boiled or eaten along with chutney.
16	*Cannabis sativa* L. True hemp	*Ganja*	*Ganja*	Leaf – Fried with flour and eaten as flour cake.
17	*Capsicum annum* Var. fasiculatum Baily. Bhut jolokia/Red Naga chilly	*Masi arokpa*	*U-Morok*	Fruit – Fresh chilly eaten as chutney and dry ones added in curry.
18	*Carica papaya* L. Papaya	*Awathapi*	*Awathabi*	Fruit – Plain boiled or eaten as simple cooked curry or raw fruit eaten mixed with chutney.

19	*Centella asiatica* Linnaeus. Urb. Indian penny wort	*Anleiphon/ Ariphon-*	*Peruk*	Whole plant – Plain boiled or eaten mixed with chutney.
20	*Cinnamomum tamala* (Buch-Ham) Indian bay leaf/Tajbat	*Tastepat*	*Tastepat*	Leaf – Added in meat curry and tea.
21	*Cissus adnata* Roxb. [Vitaceae]	*Kokngoujen*	*Kokngouyen angouba*	Leaf - Fresh leaves of *Cissus adnata* are eaten as simple boiled curry with dry fish/meat.
22	*Citrus × aurantifolia* Linn. Burm. Common/Sour Lime	*Champra*	*Chambra*	Fruit – Eaten freshly or after food.
23	*Citrus maxima* (Burm.) Merr. Pomela/Citrus grandis	*Nobab*	*Nobab*	Fruit – Ripe fruit eaten as fruit.
24	*Clerodendrum colebrookianum* Linn. Moon. Bleeding heart	*Anphui (Soknoudon)*	*Moirang Khanum*	Leaf – Plain boiled or cooked with pork curry.
25	*Curcuma caesia* Roxb. Black turmeric/Black zedoary	*Aisan ahangpa*	*Yaimu*	Rhizome – Added little in fried curry or eaten raw early morning.
26	*Curcuma longa* Linn. Turmeric	*Aisan*	*Yaingang*	Rhizome - Added in curry and some extracted juice before food.
27	*Dichrocephala integrifolia* (L.f) Kuntze. Dicrocefala, Phakchee doi	*Aralu*	*Lalukok*	Whole plant – Cooked as simple curry with fishes.
28	*Elatostema lineolatum* Wight. Lined Elatostema, Ediew	*Tuidaitoh*		Leaf – Plain boiled or eaten as stew with rice and meat.
29	*Elsholtzia griffithii* Lengser Rep, Lengmesel	*Lengtu*	*Lomba*	Leaf and flowers – Added in yam curry and chutney.
30	*Elsholtzia ciliata* (Thunb.) Hyland Vietnamese balm	*Areng*	*Teckta*	Leaf and flower – Added in yam and meat curry.
31	*Eryngium foetidum* Linn. Saw tooth	*Awa-neem (Somey)*	*Awa-phadigom*	Leaf – Added in beef and pork curry and in chutney.
32	*Eupatorium cammanoi* Linn. Hempagrimony	*Aripung/ Renglei*	*Langthrei*	Leaf – Eaten raw when stomach upset.
33	*Ficus tsjahela* Burm. Hook.f. Pilkhan, South Indian fig.	*Abung*	*Khongnang tarung*	Leaf – Fresh tender leaves eaten as plain curry or as chutney.
34	*Gynura cusimbua* D. Don/S. Moore. Hill gynura	*Muhun lou*	*Tera paibi*	Leaf – Plain boiled or eaten along with chutney.
35	*Hedychium flavum* Roxb. Ginger lily	*Soklei/ Iphung*	*Loklei*	Stem (young shoot) – Plain boiled and eaten mixed with chutney (*Eromba*)
36	*Hibiscus acetosella Welw. ex.Hiern.* Cranberry hibiscus	*Anthui ashangpa*	*Sougree anganba*	Leaf/Fruit – Plain boiled or eaten as simple cooked curry with dried fishes.

37	*Hibiscus cannabinus* Linn. Green/Red sorrel	*Anthui*	*Sougree (ashinba)*	Leaf and seed cover – Plain boiled or eaten as simple cooked curry with dried fishes. Dried seed covers are boiled and taken as red tea.
38	*Houttuynia cordata* Thunb. Lizard tail, Heartleaf	*Anleithaang*	*Toningkok*	Leaf and stem – Eaten raw along with chutney.
39	*Ipomoea batatas* (Linn.) Lam. Sweet potato	*Alu Mangkara*	*Mangara (Alu)*	Leaf – Plain boiled or eaten along with chutney.
40	*Justicia adhatoda* Linn. Adulsa/Malabar nut	*Anha (Chikpa)*	*Nongmankha*	Leaf and flowers – Cooked simple with dried fishes or fried with pork fats.
41	*Leucas aspera* (Willd.) Linn. Thumbai	*Wanchei ann*	*Mayang yenbum*	Whole plant – Cooked simple with dried fishes.
42	*Litsea cubeba* (Lour) Pers. Mountain pepper, Litsea	*Singngathai*	*Usingsha mapan*	Fruit, Flower – Fresh flower and fruits eaten along with chutney.
43	*Lysinachia candida* Lindl. Manipur loosestrife	*Kengoi*	*Kengoi*	Whole plant – Cooked as simple curry with dried fishes.
44	*Menthe arvensis* Linn. Spearmint/Pudina	*Nungsi hidak*	*Nungsi hidak*	Leaf – Eaten along with chutney or with sour fruits.
45	*Meyna laxiflora* Robyns. Muyna/Helu/Alu	*Theipi*	*Heibi*	Leaf and dry fruit – Fresh leaves eaten mixed with chutney made with dried pea powder.
46	*Musa sapientum* Linn. Banana stem/Pseudostem and flower	*Changlong*	*Laphu*	Stem and Flower – Boiled and eaten as chutney (Eromba) or cooked along with pork curry.
47	*Oenanthe javanica* Blume. DC. Java water dropwort/Water celery	*Komprek*	*Kongprek*	Leaf – Plain boiled or eaten mixed with chutney or as simple curry with fishes.
48	*Oroxylum indicum* Linn. Kurz. Indian trumpet flower	*Maklong*	*Shamba*	Fruit – Fresh fruit roasted and eaten along with chutney.
49	*Parkia timoriana* (DC.) Merr. Tree bean (Syn. Parkia timoriana)	*Jongtak*	*Yongchak*	Fruit/Seed – Fresh fruit eaten as chutney or fried or as *Eromba*.
50	*Phaseolus lunatus* Linn. Butter bean, Lima bean	*Kalandri*	*Kalandri*	Leaf – Plain boiled or eaten as simple cooked curry with fishes and meat.

51	*Plantago major* Linn. Broad plantain	*Antapot*	*Yampat*	Leaf/stem – Plain boiled.
52	*Polygonum odoratum* Lour. Smartweed/Vietnamese mint	*Phakphai*	*Phakpai*	Leaf – Added fresh leaves in chutney (Eromba)
53	*Rhus succedanea* Linn. Wax tree/Wild varnish dry galls	*Hokma*	*Heimang*	Leaf – Eaten along with chutney or sour fruits.
54	*Scutellaria discolor* Colebr. Skullcap, Daamapaate/Nilo	*Ann lakhat*	*Ching yenakha*	Whole plant – Plain boiled with little common salt.
55	*Solanum indicum* Linn small Yellow berried nightshade	*Samtrok arikpa*	*Leipungkhang*	Fruit – Boiled or fresh eaten mixed with chutney.
56	*Tamarindus indicus* (L.) Tamarind	*Mangke*	*Mangge*	Fruit – Matured or ripe fruit eaten with common salt.
57	*Trachyspermum roxburghianum* DC. Craib. Wild celery/Ajmod	*Ching Pathikhom*	*Ching pathikhom*	Leaf/whole plant – Added in curry and chutney.
58	*Viola pilosa* Blume. Viola, Banafshah	*Shantang leipak*	*Huikhong*	Leaf – Cooked simple with dried fishes and meat.
59	*Zanthoxylum acanthopodium* Pricky winged leaf/Sheguan pepper	*Singree*	*Mukthrubi*	Leaf and seed – Added in dog meat, beef and pork curry and chutney.
60	*Zingiber officinale* Rosc. Ginger	*Aithing*	*Shing*	Rhizome – Added in fish and meat curry.

Chapter 4

Chothe Indigenous Medicines

Introduction

There are also certain ethno-botanical plants in which the Chothes used as ethno-religious medicinal plants in healing and curing various sicknesses, ailments and diseases or to ward off from evil spirits. These herbs and plants are used especially, to fight against air and water borne diseases and sicknesses like sudden fever, stomach ache, swelling body parts and so on that are common among the tribal people. They believed that if a person had come in contact with certain types of evil spirits, or deconsecrated their places, or abuses them in different ways, which is why, the spirits are annoyed and angry, and punishes the person with certain types of sicknesses, causing them trouble until they repent. Therefore, with such beliefs, the Chothes and cognates tribes of Manipur like Aimol, Chiru, Maring uses these herbs and plants to ward off against the evil sprits from harming them especially on their journey.

There are also objects of animal's part like Lion's tooth, Tiger's tooth, Bear's tooth, Thunder-bolt stones and others that are used in similar way to ward of the evil spirits from causing harm and inflictions. These objects do not involve magico-religious rituals and sacrifices. But this is what their forefathers believe and is practice by some people even today that the evil forces or spirits avoids coming close to persons wearing such animal's objects. Some priest even said that the spirits are scared or dare to come close to persons carrying a gun or a bullet.

Table 9: Chothe Ethnomedicinal Plants with Religious Significances

Sl. No.	*Botanical Name (Family)- [Common Name]*	*Local Names – Chothe/ [Manipuri]*	*Part of Plant Used*	*Mythical Beliefs and their Usage for Religious Purposes*
1	*Allium sativum* Linn. [Alliaceae] Garlic	*Satun/ [Chanam]*	Bulb	Dispeller of evil spirits - The typical strong aromatic smell is belief to dispel the evil spirits from inflicting harms or getting close to humans. Therefore, it hangs in front of the main door or kept under the mattress to avoid nightmares, bad dreams and incubus.
2	*Cassia timoriensis* Wlld. [Caesalpiniaceae] Stout prickly climber	Laikee [Laikee]	Stem	Ward of evil spirits - The core of the dried stem is used as beads around the waist or neck to ward of evil spirit especially, in children.
3	*Curcuma caesia* Roxb. [Zingiberaceae] Black turmeric/Black zedoary	*Aisan ahang/ [Yaimu]*	Rhizome	Dispeller of Magical Spell - Some local medicine man/priest asked a person to drink the extracted crushed liquid of this rhizome mixed with little water and salt after offering a ritual when considered suffering from the magical spell and charms. One may also have little piece of it daily for 7 days as anti-dote to nullify the magical charms items eaten by a person.
4	*Curcuma longa* Linn. [Zingiberaceae] Turmeric	*Aisan/ [Yaingang]*	Rhizome	Divination and Evil Spirit Dispeller - It is used in specific village rituals. Mythically the colour and smell is believed to have power to dispel the evil spirit spell so it is apply by a person on the forehead while on strange journey (esp. night time) or when having bad dreams and nightmares.
5	*Curcuma zedoaria* (Christm.) Roscoe. [Zingiberaceae] White turmeric, Kentjur.	*Aithing angoupa* *[Sing angouba]*	Rhizome	Avoid nerves constrain, Stomach ache (To ward off evil spirit) :- Chew a small piece of the rhizome early in the morning to relief from stomach ache and to ward off from evil eyes and evil spirits so that nothing happens to you like stomach ache or head ache on the journey.
6	*Cynodon dactylon* Pers. L. [Poaceae] Doob/Bermuda-grass	*Tingthou/ Bangru-bange/ Sanahuplang* *[Tingthou]*	Leaf [Raw]	Children stomach problem - When children are believed to suffer from sudden stomach ache and indigestion the fresh leaves of *Cynodon dactylon* is crush and the extracted liquid mix with little water is given. The belief is that the child is suffering from the air born disease of evil python's breath.

Sl. No.	*Botanical Name (Family)- [Common Name]*	*Local Names – Chothe/ [Manipuri]*	*Part of Plant Used*	*Mythical Beliefs and their Usage for Religious Purposes*
7	*Erythrina variegate (alba)* Linn. [Fabaceae] Indian coral tree, Lenten tree, Tiger's claw.	*Hungthu angoupa* *[Kurao angouba]*	Bark	Dispeller of evil spirit - The bark of a stem or branch is made into small beads and worn around the child's neck or as arm band or as waist band to protect from evil forces. One will notice many Meitei's (Manipuris) wore it inside the *chandra* (copper beads). Traditionally, in the past they used to hand the spiky branch in front of the door or house to ward of from any evil forces.
8	*Eupatorium cammanoi* Linn. [Asteraceae] Hempagrimony	*Ariphung/ Renglei/ [Langtharei]*	Leaf	Personal and Family Rituals - The Chothe used three or five or seven or nine leaf twigs for certain types of ritualistic offerings, especially related to water and air borne sickness with a belief to appease the evil spirits, that the plant is God's gift.
9	*Goniothalamus sesquipedalis* (Wall.) Hook. F. & Thomson. [Annonaceae] Goniothalamus	*Aham* *[Leikham]*	Dry leaf	Evil Spirit Dispeller - The local believes that the aromatic scent of the burnt dried leaf of *Goniothalamus sesquipedalis* ward of any kind of evil spirits that follows a person especially during evening and night time movements out the house that tries to cause infliction. Therefore, usually people burnt the dried leaf and the scent is spread around the person and inside the house so that the evil spirit is cast out by its strong repelling smell. Sometimes the ash is applied on the forehead before taking a night journey or strange places. In case of fever, sometimes the ash is mix with some water and drank to free from feverish problems of possession.
10.	*Hedychium coronarium* (J.Koening). [Zingiberaceae] White garland lily, *(Zingiber cassumunar* Roxb.)	*Aidou [Yaingang ariba/* *Tekhal yaikhu]*	Rhizome	Ward of evil spirits esp. babies - The specie of *Hedychium coronarium* is a very important herb for Chothe in child's naming ceremony. Traditionally, the rhizome is made into beads and made to wear around the child's neck as necklace to protect from any kind of sicknesses and to ward off from evil forces that tries to torment the child.
11.	*Holskioldia sanguine* Retz. [Lamiaceae] Chinese hat plant, Cup and Saucer	*Leiningthung asanpa* [*Kharam leisok anganba*]	Leaf	Infertile woman - The young married woman offers this flower to their deity with a belief that she will be able to conceive a child which she could not for certain reasons. The belief is connected to the history of Moirang King Chothe Thangwai Pakhangba and Sunulembi his beloved wife.

Sl. No.	*Botanical Name (Family)- [Common Name]*	*Local Names – Chothe/ [Manipuri]*	*Part of Plant Used*	*Mythical Beliefs and their Usage for Religious Purposes*
12.	*Meyna laxiflora* Robyns. [Rubiaceae] Muyna/Helu/Alu	*Theipi* [*Heibi*]	Leaf	Village Ritual - Traditionally, the Chothes used the leaves as an essential item along with local yeast in preparing chutney for evil spirit in certain village rituals.
13.	*Ocimum sanctum* Linn. [Lamiaceae] Holy basil	*Tulsi* [*Tulsi*]	Leaf	Sacred plant – Because of its highly medicinal value, some Chothes and Meitei's planted this herb as sacred plant in their courtyards.
14.	*Oroxylum indicum* Linn. Kurz. [Bignoniaceae] Indian trumpet flower	*Maklong* [*Samba*]	Bark, fruit and stem	Protects from evil forces - The *Oroxylum indicum* tree of white flower is useful in many ways and has religious implication. They belief that a python can be kill only by placing it head on the stem of this plant when its day is fixed to die. Therefore, the plant is planted as valuable plant in the courtyard.
15.	*Toona ciliate* M. Roem. [Meliaceae] Red cedar	*Atithing* *[Tairen]*	Leaf	Dispeller of evil spirits - Traditionally, few stalks of leaves were used in various religious ceremonies to ward of evil spirits from harm. The leaves are used to sprinkle holy water from a bowl or mug or bucket of water or like in house purification ceremony.
16.	*Zingiber officinale* [Zingiberaceae] Ginger	*Aithing* *[Shing]*	Rhizome	Evil Spirit Divination - Besides using it as culinary item, the Chothe used in certain village rituals in divination. Mythically they believed that the strong pungent smell and taste acts as antidotes for certain poisonous things eaten or repellent of evil spirits.

B. Chothe Indigenous Medicines

In the past, traditionally the Chothe and other cognate tribes believed and practiced certain aspects in healing, curing and accumulating wealth. Some of these aspects are:

1. **Thunderbolt Stone:** Air and water born illnesses - In case of sudden fever and stomach ache the thunderbolt stone of water nature is smeared on a hard-clean stone and the extracted liquid is collected by cotton in a glass where it is mixed with little water and applies on the whole body and one is also asked to sip some repel the antioxidant substances. Perhaps, if they believed that the person suffers from evil eyes or was caused by evil spirits the brownish thunderbolt stone (fire nature) is used. This thunderbolt stone has religious significance in connection with the mythical beliefs. Such genuine thunder-bolt stone are scientifically believed to contain certain amounts of unknown elements considering part of the meteorites. The Chothe called it as *Tecklei* and the Meitei as *Braja nung*. There are two main types – Greenish stone is associated with water character and the brownish stone with fire character.
2. **Rhino horn**: Cancer and tumour - The horn is smear on a hard stone and the extracted liquid is collected with cotton and is mixed along with other plant's item in the preparation of ethnomedicines.
3. **Fermented Rice Beer Residue:** Sprain, fracture - In earlier days, if a person got sprain or fracture on his leg or hand the fresh fermented rice beer residue (*i.e.* the fermented rice) is wrapped around the sprain or fracture part for 2-3 days. This process is repeated with the new fermented rice residue till it is heal. Sometimes, if the injury is very serious some fresh turmeric is crushed and mix along with it and the paste is applied by bandaging around the injured area.
4. **Lion's/Tiger's tooth:** Cancer, Tumour – The tooth is smear on a hard stone and the extracted liquid is collected and mixed along with other ethnomedicinal plant's items.
5. **Tiger's tooth:** Ward of evil spirits – Some wore around the neck to ward of evil spirits from contacts on their journey.
6. **Bear's bile**: Measles, small pox – Small amount of the dried bear's bile is mixed with water and given in small dosages twice a day for 2-3 days to person infected with measles and small pox.
7. **Pig's bile**: Measles, small pox - Small amount of the dried bear's bile is mixed with water and given in small dosages twice a day for 2-3 days to person infected with measles and small pox.
8. **Python's bile**: Measles, small pox - Small amount of the dried bear's bile is mixed with water and given in small dosages twice a day for 2-3 days to person infected with measles and small pox.
9. **Giant water bug**: Snake bites – In case a person is bitten by a snake, one is suggested to paste the head (even if it is dried) on the bitten part to

automatically ooze out the poison. This is in connection with the belief that giant water bug can killed a snake by its sting.

10. **Python's fats**: Burn - Immediately apply the python's fat on the burn areas to sooth the hotness and removed the burnt marks later.
11. **Python's eye:** Lucky charm - Used as lucky charm by card players with a belief that they can visualised the other's cards and win the game.
12. **Twin king chilly:** Lucky charm - Card players carry hiddenly such twin king chilly (Naga chilly/Umorok) as lucky charm to win the bet.
13. **Saliva:** Insect's bites - In case of insect bite, the Chothe applies spit (Saliva) on the bitten part. They also apply a small amount of lime (lime stone) in case of allergies on the specific body areas or on the lower tip of the ear. Some even apply the residue of burnt kerosene in the areas of itchiness or allergy.
14. **Sitamasi (Misri)**: Mixture – Some amount of crystal cane sugar is often added in preparation of concoction or decoction and herbal syrup ethnomedicines by the locals especially when the herbal substances are extremely bitter or sour or have very pungent or clumsy in taste. It is added either to reduce the bitterness, sourness or very pungent taste.
15. **Honey**: Mixture – A tea spoonful or more is often added in preparation of ethnomedicine syrup by the locals.
16. **Wasp mud**: Mum – The mud of the house wasp is collected and mix with little water or saliva and is apply regularly 3 times a day on the mum area for a 3-5 days. It is believed to act as a cooling agent and heals the mum.
17. **Outer's paw**: Fish bone stuck on throat – When a person ate fishes and a fish bone perhaps stuck on the throat, then the Chothes rubs down the throat with the outer's paw saying three times its name as *'aram, aram, aram'* where the fish bone was stuck. Keep repeating 3 times till the fish bone disappeared.
18. **Fowl's feather**: Worm in babies - Burn the fowl good feathers and collect the ash in cotton and add little honey and put around the child's anus at night time before the child slept. After 2-3 hours, if one inspects the anus of the child while sleeping one may find the worms around the anus trying to escape. Catch it with forceps and kill them.
19. **Rice**: Dog's bite and Blights in the tongue - Chew the raw rice and apply the paste on the dog's bitten part and place a silver coin on it. If there is any blights or prickly thing develops on the tongue chew some raw rice let the rice starch heal the prickly. Repeat 3-4 times a day.
20. ***Pharekbu** (Tambak yenbung* in Manipuri and *Janganlek* in Rongmei*)*: Anaemia, Irregular menstrual problem (Accumulation of bad blood) - Women suffering from anaemia (lack of red blood) and irregular menstrual problem who cannot gain weight may simply boiled and eat the fresh leaves as simple cooked curry along with fresh local cat fish (*Ngakara*) twice a day for 5-7 days. This will extract the bad blood concentrated in the woman's body and enhance fresh blood.

21. ***Asam arengpu*** (*Upupa epops*/Hoopoes) in Meitei *Chonga laba*: Stone cases – Person suffering from liver, kidney and bile stone cases may boil some meat of the Hoopoes meat with some fresh leaves of *Cissus adnata* Roxb. (*Kokngouyen* in Meitei) and served the soup every day for a week. The bird *Upupa epops* Leach 1820, **belong to the family of** Upupidae and is **common known as** Hoopoes.

C. Pictures of Chothe Indigenous Medicines

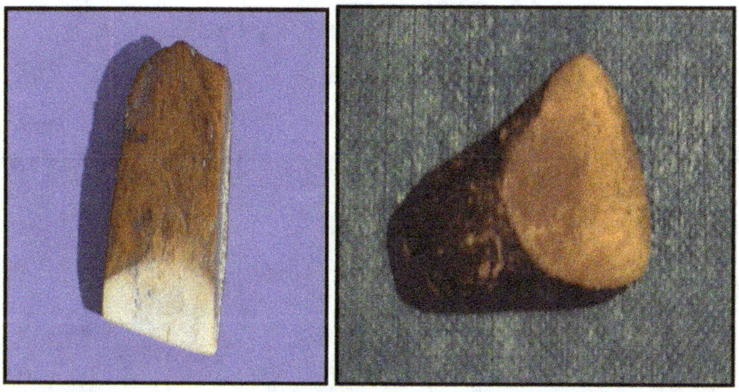

Thunder bolt stones (*Tecklei or Braja nung*)

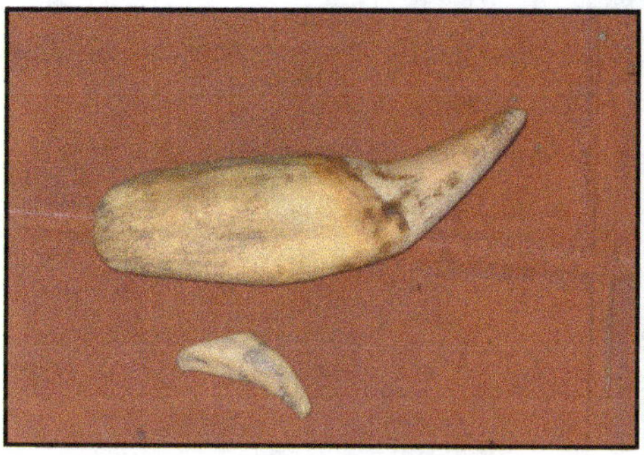

Tiger's tooth (*Akei ha*)

Giant water bug (*Nausek arokpa*)

Pig's dried bile (*Amit kha/Masing kha*)

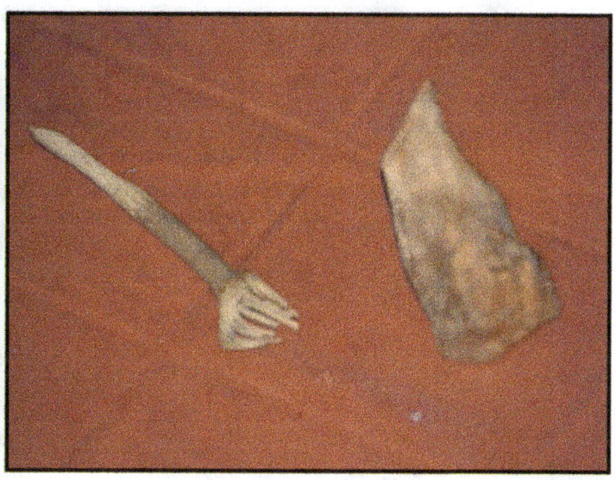

Laikee/Laikee's dried stem

Mithun's horn

Fowl's Feather (*Ai mui/Ar mui*)

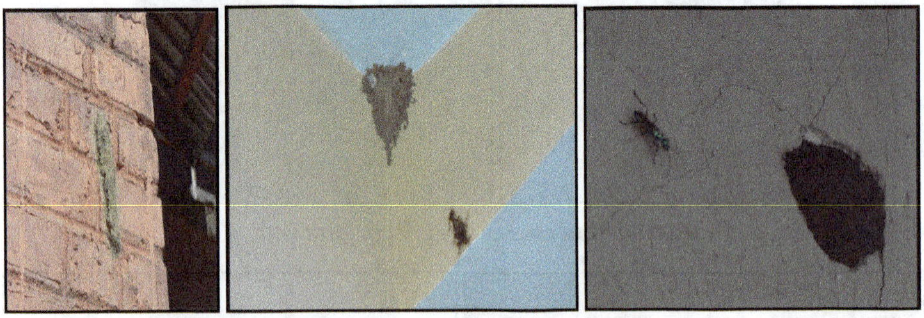

Wasp House (Hui arei inn sapi)

Rice (*Chisai/Cheng*)

Crystal cane sugar (*Sitamasi/Misri*)

Honey (*Hui tui*)

Outer leg (*Aram kae*)

Kaang and Soapy Seed

Traditional fermented rice beer process style

Decoction solution

Local medicine man's ethnomedicinal items

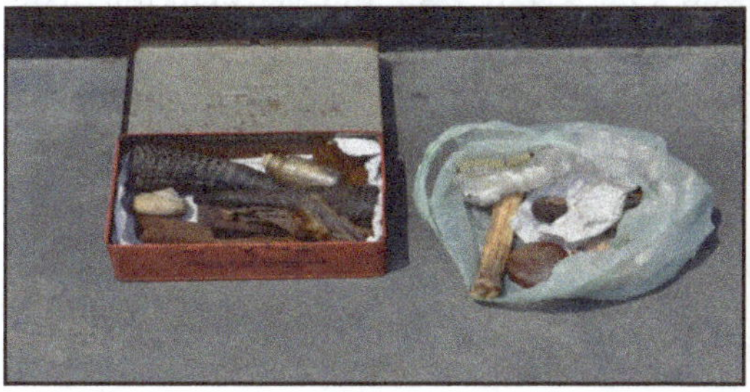

Ethnomedicinal items storage style

Hoopoe bird (*Asam arengpu*)

Local tablets made of gooseberry with other ingredients.

Traditional distillation process of alcohol (Liquor)

Botanical Name – (Unknown),
[Pharekbu in Chothe, Tambak yenbung in Manipuri and Janganlek in Rongmei]

Diseases/Treatment: Anaemia, Irregular menstrual problem (Accumulation of bad blood).

Women suffering from anaemia (lack of red blood) and irregular menstrual problem who cannot gain weight may simply boiled and eat the fresh leaves as simple cooked curry along with fresh local cat fish (*Ngakara*) twice a day for 5-7 days. This will extract the bad blood concentrated in the woman's body and enhance fresh blood. This herb grows wild in the open grasslands. The leaves are pale yellow in colour, hairy and oval in shaped with often four leaves. The flowers are tiny yellowish bloom in cluster from a long stalk from the centre of the plant. The leaves taste simple with little aroma when cooked. The Chothe and other tribes gathers from forest and eaten as simple cooked curry with fishes or as chutney.

Chapter 5

Conclusion

Discussion and Results

The Chothes have used various ethno-botanical plants such as herbs, creepers, climbers, shrubs, trees, and grasses or weeds in the treatments of healing and curing various ailments, sicknesses and diseases using the plant's parts like leaves, barks, seeds, roots, flowers, stems, rhizomes. Though some used single plant part but sometimes they used compound of two or more plants parts, and even adding either honey or crystal cane sugar (*Sitamasi/Misri*) in the preparation methods. These plant parts are used in diverse forms of fresh or raw, dried or powder forms, or as culinary or concoction/decoction, or as syrup and tonic form, and externally as ointments and balms. These are the traditional methods of treatment used by most tribal peoples of the world. Such medications are usually prescribed by local medicine man or priest or priestess or village elders who have acquired their knowledge through their experiences, trial and error methods considering effective. It is also found that in the study certain same species of the plant was used differently by different ethnic groups or individual depending upon their experiences and belief[1]. These ethno-botanical plants are grown naturally in wild or cultivated in most natural fertile soil absolutely free from artificial fertilizers, pesticides and toxic substances, and therefore, one can assumed that these plants are one of the best and highly rich in each original nutrient state since found in their natural habitats. The results of the study among the Chothes are discussed herein.

The Table 1 shows that there are 130 identified ethno-botanical plants that belong to 58 families (See Table 2) that are used in the treatment of more than 69 ailments and diseases (See Table 3) among the Chothe tribe of Manipur, Northeast India. Most of these plants are found in their surrounding environments.

The Table 2 indicates that out of 130 ethno-botanical plants 58 (fifty-eight) belongs to different plant families. The most common 15 families with highest frequencies are: i) Lamiaceae – 11, ii) Zingiberaceae – 9, iii) Fabaceae – 8, iv) Asteraceae – 7,

v) Apiaceae – 4, vi) Malvaceae – 4, vii) Phyllanthaceae – 4, viii) Rubiaceae – 4, ix) Solanaceae – 4, x) Amaranthaceae – 3, xi) Araceae – 3, xii) Convulvulaceae – 3, xiii) Eupharbiaceae - 3, xiv) Meliaceae – 3, and xv) Poaceae – 3. The rest of the families shared a frequency of 1 and 2 each.

Table 3 shows that there are about 69 ailments, illnesses, sicknesses and diseases of various categories were treated by these 130 identified ethno-botanical plants as ethnomedicines by the Chothe tribe of Manipur. Some of the most common ailments, sicknesses and diseases prevail among them are: i) Cold and Cough – 11, ii) Fever – 9, iii) Boil – 9, iv) Blood purification/Anaemia/Enhance blood – 9, v) Kidney failure/Kidney Stone cases/Urinary infection problems – 8, vi) Diabetes – 8, vii) Gastritis – 7, viii) Jaundice – 7, ix) Cuts and wounds – 7, x) Irregular menstrual problems – 6, xi) Chronic ulcer – 6, xii) Dysentery – 6, xiii) High Blood pressure – 6, xiv) Piles – 6, xv) Hypertension - 5, xvi) To enhance body immune systems – 5, xvii) Asthma – 5, xviii) Worms infestation – 5, xix) Constipation – 5 and, xx) Diarrhoea - 4. The rest of the ailments and diseases or sicknesses shared the frequency of 4, 3, 2 and 1 each.

Out of 69 different ailments, sicknesses and diseases certain illnesses are considered serious by the Chothes (see Table 1 & 3). Such as: irregularly menstrual problems of women, urinary infection or kidney stones cases, diabetes, cancer/tumour, jaundice, chronic ulcer, anaemia/blood purification to enhance blood especially for women and migraines especially for married women. The rest of the ailments provided in the Table 3 are considered less danger as they believed it can be easily treated and posed no serious threat to life. A similar study conducted by the author among the Maring tribe of Manipur found that some local medicine men are experimenting with their ethnomedicines on breast, cervical cancer and paralysis claiming that they can cure it (*Ethnomedicinal plants of the Maring tribe of Manipur, Northeast India* (2018).

However, the Chothes considered measles among the children a serious illness. Therefore, a small amount of 1 ml of dried bear's bile or pig's bile mixed with honey is given twice a day for 3-5 days or till cure. Sometimes the fruit of *Solanum indicum* is crush and mix with honey is given to children and babies in small dosage of about 2 - 3ml twice a day for 3 days or more, so that the measles from inner is repelled outside.

Mr. M. Kundo of Lamlanghupi said that "Depending upon the seriousness of the illness and age of the person the traditional indigenous medicines are prepared. If a person had developed to a second-degree infection the ethnomedicines are prepared in concentrated form. However, if the patient is a child, the dosages is minimise compared to adults from 3 tea spoonful to ½ or 1 tea spoonful."

However, every traditional treatment does not mean that it proves effective to all the patient while some may not due to certain reason. Mr. Th. Sirochong of Lamlanghupi said that "Sometimes certain ethnomedicine may prove not effective to all because of his/her immune system or his/her blood group resistancy. In such case, another variety of ethnomedicine may be used to kill the virus, bacteria and germs or subdued the virus present in his body."

The Chothe used different types of plants which are in different forms and size and nature of the plants. Some are trees, shrubs, sub-shrubs, herbs, creepers and climbers. The Table 4 *Forms of the Plants* indicates five categories of plants. They are: i) Tree – 22, ii) Shrub – 34, iii) Herbs – 59, iv) Creeper – 9 and, v) Climber – 6. The maximum frequency is shared by herbs which show that the Chothes used mostly herbs as their ethnomedicinal plants to heal and cure various ailments, sicknesses and diseases. Some of the trees like *Chingthrao* (sour leaf variety) are very rare and found only in the deep interior forest of neighbouring villages.

The Table 5 *Types of the Plants* indicates two categories of plants. They are: i) Seasonal – 99 and ii) Perennial/Annual – 31. Most of these 130 identified ethno-botanical plants are seasonal based as they grow during their growing season. Therefore, certain useful and rare plants are dried and preserved for future used.

The Table 6 *Parts of the Plant Used* indicates eleven categories. They are: i) Leaf – 84, ii) Fruit – 19, iii) Whole plant – 15, iv) Seed – 5, v) Bark – 6, vi) Stem – 13, vii) Root – 6, viii) Rhizome – 8, ix) Flower – 6, x) Bulb – 1 and, xi) Tuber – 1 respectively. The leaf shared the maximum frequency compared to other parts of the plants.

Table 8 *Treatment Methods* indicates three categories. Such as: i) Oral – 90, ii) External – 49 and, iii) Both cases – 9. The Chothes are seen administering these ethno-medicinal plants in three forms as orally, externally and in both forms.

The administration of the ethnomedicine depends upon the nature of ailments and sicknesses either orally or externally. Orally are administered either as tonic, concoction, decoction, or eaten as culinary items, especially the fresh leaf or part of the plant. These ethno-botanical plants and herbs are considered harmless and eatable. They considered that it rather provides the required minerals and vitamins to the body systems that is lacking or suffering from deficiencies or malnutrition, thereby enhancing the body immune systems.

Externally means the treatment are administered as externally on body parts as ointment, balm, paste for the cases of cuts, burn, sprain, fracture, septic, *etc.* However, certain ethno-botanical plants and herbs are either used for both cases as orally or externally depending on the types of ailments and sicknesses.

However, in certain respect compound items are also used in preparation of ethnomedicine for multi-purpose treatment like for cough, dry cough and inflammation of the stomach by the Chothes. For example, simply boil in about 2 litres of water some leaves, (seeds if leaf is not available) and rhizome of *Centella asiatica, Justicia adhatoda, Eryngium foetidum, Zanthoxylum acanthopodium* and *Zingiber officinale* by adding little common salt and served about half a glass of the decoction solution 3 times a day for 3-5 days or till relief of the cough. It is suggested that for some people it is very good tonic or syrup for dry cough, cough and inflammation of stomach and even constipation.

The significant of all these Chothe ethno-medicinal plants and herbs is that whether it is by taste, look, or smell, generally it has its own unique qualities. Some have pleasant or strong aromatic smell, some are attractive in looks, some has sweet flavour, some slightly bitter to extremely bitter in taste, some are little pungent while others have no taste at all. Some are available at the courtyard itself like weeds, while others are grown in the thick forest areas only. Most of these plants are seasonal and bloom flowers for very short period. The non-eatable or considered poisonous plants are especially used for external purposes only. These non-eatable or poisonous plants taste deadly and appeared less appealing in looks or creepy.

The Chothes consumed certain herbal plants as vegetables or prepares in simple indigenous food items by simply boiling or simple cooked curry considering eatable and good for health, often with or without knowing the medicinal properties and values. They learnt that some plant's leaf/seed/rhizome gives better taste when cooked with specific items like with dry fish or meat, or when mixed with certain appropriate ingredients. For example, the leaf of *Awaneem* (*Eryngium foetidum*) is added in fresh beef and pork curry to enhance the taste and aroma, or *Singree* (*Zanthoxylum acanthopodium)* in fresh dog meat, beef and pork curry to reduce the foul smell. Both these two plants leaves are also added in preparing dog's and pig's sausages which is a delicacy for the people.

Their simple traditional local food habit is believed to have kept them healthy and strong, and free from various ailments in the past because what they consumed are natural products from their gardens, courtyard and forest which are not contaminated with fertilizers, pesticides, germicides and other poisonous toxic substances. It is also noticed that most of these plants are commonly eaten by the North-Eastern and Asian communities. However, nowadays the Chothe have changed their food habits from simple cooked to fried curry because of the impact of westernisation and acculturation with the plain peoples besides, other media cookery influences.

Table 8 indicates that there are about 60 (Sixty) identified ethno-botanical plants out of 130 that are taken as food habits by the Chothes in different forms of culinary as – simple boiled, simple cooked curry, chutney or along with local made chutney (*Eromba*) besides, taken as fruits. These items are consumed occasionally during the season and some are preserved in dry form and taken when required.

However, in Chapter Four, the Table 9 indicates that there are about 16 identified ethno-botanical plants that has religious significances. The Chothes believed that these plants helped in curing, healing and protection from certain kinds of sicknesses and illnesses. Therefore, some are still preserved and used by them when require and demanded. They are basically used to ward of the evil spirits from contacts, nightmares, bad dreams, protection from certain sicknesses or diseases especially among babies and children against viral infections.

The Chothes also used these plant's parts like ginger, turmeric, *Goniothalamus sesquipedalis*, *Meyna laxiflora*, and *Toonna ciliate* in divinations and rituals of personal, family and village offerings. Out of 16 plants the most common families are – Zingiberaceae – 4 and Lamiaceae – 2, while the rest shared frequency of one each. The plants that has religious significances are given below:

1. *Allium sativum* L. (Garlic- *Satun/Chanam*)
2. *Cassia timoriensis* Wlld. (Stout prickly climber- *Laikee)*
3. *Curcuma caesia* Roxb. (Black zedoary – *Yaimu*)
4. *Curcuma longa* Linn. (Turmeric – *Aishan/Yaingang*)
5. *Curcuma zedoaria* (Christm.) Roscoe. (White turmeric – *Yaingang agnouba)*

6. *Cynodon dactylon* Pers. L. (Bermuda-grass – *Tingthou*)
7. *Erythrina variegate (alba)* Linn. (Indian coral tree – *Hungthu toh/Kurao*)
8. *Eupatorium cammanoi* Linn. (Hempagrimony- *Renglei/Langtharei*)
9. *Goniothalamus sesquipedalis* (Wall.) Hook. (Goniothalamus – *Aham/Leikham*)
10. *Hedychium coronarium* J. Koening. (White garland lily – *Aidou / Tekhal yaikhu)*
11. *Holskioldia sanguine* Retz. (Chinese hat plant – *Leiniingthung/Kharam leisok*)
12. *Meyna laxiflora* Robyns. (Muyna – *Theipi/Heibi*)
13. *Ocimum sanctum* Linn. (Holy basil – *Tulsi*)
14. *Oroxylum indicum* Linn. Kurz. (Indian trumpet flower – *Maklong/Shamba*)
15. *Toona ciliate* M. Roem. (Red cedar- *Atithing/Tairen*) and
16. *Zingiber officinale* Linn. (Ginger – *Aithing/Sing*).

We also notice that a section in Chapter Four - *Indigenous Ethnomedicines of the Chothe,* there are about 21 items that are animal's part, insects, objects and substances that are used by the Chothes in curing and healing certain ailments, sicknesses and diseases besides, the provided ethno-botanical plants herein. For example, the thunder-bolt stone that the local medicine man and priests preserved are of different colour, often like an axe shaped. Some are greyish, greenish and brownish in colour. The stones are believed to be meteorite stones since, they are very hard, smooth and different from the ordinary stones. When a person suddenly falls sick with fever or suffers from stomach upset the thunder-bolt stone is rubbed against a hard stone and the extracted liquid is either applied on the body's part or drank little by mixing with water. This treatment method was widely practiced by them in the past. The Chothes also preserved certain talisman stones that have magnetic power. These talisman stones are of different shaped and sizes and are mostly very black in colour. They are locally called as *Shanglung* (lit. Paddy's talisman stone) or *Asha arung* (lit. Animal's talisman stone).

The dried bear's bile and pig's bile are considered good for treating measles. Taking some solution helps repel the measles that are imbedded inside the body. The horn of a Rhino was used in treating tumour and cancer related diseases by some priests. The mud of the house wasp helps cool down the heat of mum on the jaw. Lion's tooth, and tiger's tooth are worn around the neck to protect themselves against evil eyes and evil spirits. The python's fat are preserved and is applied on the burn area to cool down. Human saliva is used as saline when insects bites them. They used giant water bug's head (fresh or dried) and sometimes a silver coin when a person is bitten by a snake with a belief that placing at the bitten part removes the snake poisons. In the past, the Chothes treasure and preserved the Otter's paw with a belief that rubbing nine times against the throat where the fish bone was stuck would disappear later unknowingly.

The Chothe traditional curing and healing methods is not limited only to human beings but also extends even to animal kingdoms like cows, buffaloes, pigs, horses, domestic fowls, dogs, *etc.* when they suffered from certain illnesses, infection or injuries. However, the practice has substantially declined with the advancement of

Veterinary medicines and technologies. The traditional medication of animals have eventually decline to great extent.

Values of Traditional Ethnomedicinal Knowledge

In the study of ethnomedicines one must keep in mind that most of the priest, priestess, local medicine man, village elders and grandparents are the traditional knowledge bankers who are experts in ethnomedicines. However, they would not disclose everything especially, the treatment method for serious ailments and diseases to interviewer apart from general medicines. They usually retain their top secret and confides only to their disciple or most trusted person, that is how the traditional ethnomedicine is handed down from one generation to the next. They have clear intentions not to reveal all their expertise and talent to others which involves certain reasons. Such as sharing their secrets would reduce the power of healing as it loose its potential charm, on the other hand, people can misused the remedy methods and create jeopardy to others, and in certain cases, they believed that their superior status of respect as an expert in the field of ethnomedicine declined when knowledge are simply shared to others and in some cases, their ethnomedicines becomes nullified.

According to Pr. Roushi (87/M) a local medicine man of Ajouhu said that "In the past, if we freely shared such valuable knowledge and its secret treatment methods to others it was strongly believed that the medicinal power reduces when treated on others again. Therefore, we hardly shared such invaluable knowledge for fear of losing its potency of the plants and its application".

Similarly, N.S Jamir from the perspective of Naga tribes of Nagaland said that 'traditional knowledge is confined chiefly to the folk-healers and old-folks residing in villages. They believed the knowledge of the medicinal efficacy of plants are lost to posterity, so they do not divulge the secret in fear that their professional supremacy will be at stake and the use of medicinal potency of the plants would be weakened or nullified[2].

Rajkumari (2013) also states that amongst the Chiru tribe of Manipur the traditional knowledge system is restricted to few identified persons in the community and the knowledge is generally inherited through the oral transfer and that generally in family lineage as there is no written document[3].

Hence, it is pertinent for such reasons that they did not openly disclose or freely share their secret knowledge and expertise to younger generations. But they claimed to share only to their most trusted person like his son or disciple whom they believed can continue practicing their secret tradition of healing and curing complicated illnesses, even to the extent of killing a rival with magical-witchcraft.

Roushi, further said that "In olden days, there used to be even deadly magico-religious witchcraft competition amongst fellow village priests or local/regional medicine men or fellow neighbouring priests in their expertise of healing and curing certain unknown illnesses. The fear of this psycho-warfare were infuse upon one another so that people do not simply take advantages of one's skill and talent without prior approval".

Conclusion

In this study of ethnomedicines among the Chothes there are about 130 identified ethno-botanical plants, out of which, most of them belongs to herbal plants which are seasonal. They used more of leaf, stem and rhizome in the form of oral decoction, concoction in the treatment of various ailments and diseases. Some are eaten as food items while others are used for external purposes only. Single plants are found using for multi-sicknesses. Most of the sicknesses treated are of common ailments and diseases like fever, cough, dysentery, urinary or kidney related problems, irregular menstrual cycle problem, jaundice, diabetes and so on.

The patients often considered effective and satisfactory using these ethnomedicinal plants in certain cases. Depending upon the degree of illness, magical-religious ceremonies of sacrifices were also employed by the village priest in the healing and curing methods. However, interesting phenomenon is the increase in the belief of Christians prayers healing programmes since, the majority are recent converts.

Medicinal plants for Diabetes, Cancer, Aids, HIV, Hepatitis, *etc*. are unknown because such diseases were unknown to them before, but on experimental basis, some are using certain plants to treat these new diseases. It may be noted that besides these few plants identified, there are several other plants which are known and unknown to the Chothes, and not recorded herein.

Most importantly, changed in eco-system, deforestation, jhum farming, constrain of economy, Christianity, ignorance of people, lack of awareness of conservation and preservation, unsystematic collection and destruction of identified plants from its natural habitat, sharp decline of elderly knowledgeable persons besides, other factors have seriously posed a threat to the existing practices of ethnomedicinal plants and indigenous medicinal in North-East India.

Notes: *Some botanical names provided under the photograph/image may not be the same species provided in certain websites since, there are different varieties of the same genus ranging from three to one hundred plus varieties. Moreover, the mapping and identification of the botanical plants are still incomplete with new species discovered everyday by various eminent scientists of the world. However, the photographs provided are actual species identified and verified by the local respondents that they used as ethnomedicines.*

References

(Endnotes)

[1]Yuhlung, Cheithou Charles. 2018. *Ethnomedicinal plant of the Maring tribe of Manipur, Northeast India*. New Delhi: Regency Publication.

[2]See above, Jamir, NS. 2012.

[3]See above, Rajkumari, Ranjana. 2013.

Suggestions

1. The Government should immediately take serious steps to encourage to document and preserved in a scientific manner all the ethno-botanical plants used as medicines for all the India tribes and especially, of Manipur tribes, since many of the tribes have not been systematically and scientifically studied.
2. The Government should provide financial assistance for the identified sensitive villages and tribes of Manipur for Herbarium, to help preserved these rare lifesaving ethnomedicines for future used besides, documentation.
3. Some experts in the field of ethno-botany or ethnomedicines should be provided to assist the sensitive village or tribe in the identification, preservation and protection of the endangered plant species along with the villagers.
4. Certain sensitive plants that have not scientifically studied should be scientifically study again to find out the various constituent component and elements of the plant what causes them to be so effective in curing and healing certain ailments and diseases.
5. Various Government agencies and NGO's should work together to check and protect the environment from further degradation, ecological disturbances and over exploitation of various endangered and rare plant species from extinction. They should also assign responsibility to help preserved and protect the deterioration of eco-systems from further degradation of various flora and fauna.

References

Ansari, SA. 1991. *Manipur: Tribal demography and Socio-economic Development*. Delhi: Anil Mittal.

Basumatary, S; M Ahamed and SP Deka. 2004. Some medicinal plant leaves used by Boro (Tribal) people of Goalpara district, Assam, (*Explore*), in *Natural Product Radiance*. Vol. 3(2), March-April. http://nopr.niscair.res.in, accessed on 02-03-2014.

Buragohain, Jitu. 2011. Ethnomedicinal Plants Used by the ethnic Communities of Tinsukia District of Assam, India, in *Recent Research in Science and Technology*. Vol.3(9): 31-42. www.scholarjournals.org, accessed on 03-04-2014.

Census of India 2011. (Population census).

Chakraborty, Raja, Biplab De, N Devanna, Saikat Sen. 2012. North-East India an Ethnic Storehouse of Unexplored Medicinal Plants, in *Scholars Research Library; Journal of Natural Product and Plant Resource*. Vol.2(1), 143-152. http://scholarsresearchlibrary.com/archive.html, accessed on 03-04-2014.

Das, Ajit Kumar; BK Dutta and GD Sharma. 2008. Medicinal plants used by different tribes of Cachar district, Assam, in Indian Journal of Traditional Knowledge. Vol.7(3), July, 446-454.

Greirson, George Abraham (Ed.). 1967. *Linguistic Survey of India*. Vol.III. Tibetan-Burman Family, Part-III (Specimens of the Kuki-Chin and Burma groups). Calcutta: Superintended of Government Printing), [reprinted, Delhi: Motilal Banarsidass].

Hazarika, Ridip; Santoshkumar Singh Abujam and Bijoy Neog. 2012. Ethno Medicinal Studies of Common

Plants of Assam and Manipur, in*International Journal of Pharmaceutical & Biological Archives*. Vol.3(4), 809-815. www.ijpba.info, accessed on 05-04-2014.

Jamir, NS; Lanusunep and Narola Pongener. 2012. Medico-herbal medicine practiced by the *Naga* tribes in the state of Nagaland (India).*Indian Journal of Fundamental and Applied Life Sciences.* Vol. 2 (2) April-June, 328-333. http://www.cibtech.org/jls.htm,accessed on 02-02-2014.

Jain, Alka; S Roshnibala, PB Kanjilal, RS Singh and H Birkumar Singh. 2007. Aquatic/semi-aquatic plants used in herbal remedies in the wetlands of Manipur, Northeastern India, in *Indian Journal of Traditional Knowledge.* Vol6(2), April, 346-351.

Jaiswal, Vidhan. 2010. Culture and ethnobotany of Jaintia tribal community of Meghalaya, Northeast India: A mini review, in *Indian Journal of Traditional Knowledge.* Jan. Vol.9(1), 38-44.

Khongsai; M Saikia, SP and Kayang H. 2014. Ethno-medicinal plants used by different tribes of Arunachal Pradesh, in *Indian Journal of Traditional Knowledge.* Vol.13(2), April. 368-376.

Mao, AA; TM Hynniewta and M. Sanajaoba. 2009. Plant wealth of Northeast India with reference to ethnobotany, in *Indian Journal of Traditional Knowledge.* Vol.8(1), Jan. 96-103.

Ningombam, DS; Deshworjit Singh and Potsangbam Kumar Singh. 2014. Ethnobotanical Study of *Phologacanthus thyrsiformis* Nees: A Conserved Medicinal Plant of Manipur, Northeast India', in *International Journal of Herbal Medicine.* Vol.1(5), 10-14. www.florajournal.com, accessed on 24/04/14.

North Eastern Council Vision (2020). 2012. Ministry of Development of North Eastern Region & North Eastern Council. (NER 2020 – NEC Look East Policy. pdf). Vol.1. http://www.indianchamber.org/northeast

Pfoze, Neli Lokho; Yogendra Kumar, Bekington Myrboh. 2012. Survey and assessment of ethnomedicinal plants used in Senapati District of Manipur State, Northeast India, in *Phytopharmacology.* Vol.2(2), 285-311. *www.phytopharmajournal.com*, accessed on 02-09-2014.

Prakash, N; MA Ansari, P Punitha and PK Sharma. 2014. Indigenous traditional knowledge and usage of folk bio-medicines among Rongmei tribe of Tamenglong district of Manipur, India, in Prakash (et al.), *African Journal of Traditional Complementary and Alternative Medicine.* Vol.11(3), 239-247. http://www.researchgate.net, accessed on 02-09-2013.

Rajkumari, Ranjana; P.K. Singh, Ajit Kumar Das and B.K. Dutta. 2013. Ethnobotanical investigation of wild edible and medicinal plants used by the *Chiru* Tribe of Manipur, India, in *Pleione.* Vol.7(1), 167-174. http://www.pleione.info/, accessed on 02-09-2013.

Rai, Prabhat Kumar and H. Lalramnghinglova. 2010. Lesser known ethnomedicinal plants of Mizoram, NorthEast India: An Indo-Burma hotspot region, in *Journal of Medicinal Plants Research.* Vol.4(13), 4 July, 1301-1307. http://www.academicjournals.org/jmpr, accessed on 02-03-2014.

Ripunjoy, Sonowal. 2013. Indigenous Knowledge on the Utilization of Medicinal Plants by the Kachari Tribe of Dibrugarh District in Assam, North-East India.

International Research Journal of Biological Sciences. April Vol. 2(4), 44-50. Available online at: www.isca.in

Roy, R. C. 1936. 'Notes on the Chawte Kuki Clan', in *Man in India*. Vol.16, 135-155.

Shankar, R; Lavekar GS, Deb S, Sharma BK. 2012. Traditional healing practice and folk medicines used by Mishing community of North East India.*J Ayurveda Integr Med*. Vol.3(3): 124-9. http://www.jaim.in/text.asp

Shakespeare, J. 1912. *The Lushai-Kuki clans*. Suman Lata, Cultural Publishing House, Delhi.

Shankar, Rama and M.S. Rawat. 2013. Conservation and cultivation of threatened and high valued medicinal

plants in North East India, in *International Journal of Biodiversity and Conservation*. Vol. 5(9), September, 584-591. http://www.academicjournals.org/IJBC, accessed on 23/11/13.

Singh, B.H; Singh, P.K., Singh, S.S and Elangbam, B. 1996. Indigenous bio-folklores and practices. Its role in biodiversity conservation in Manipur, in *Journal of Hill Research*. Vol.9(2), 359-362.

Singh, H.B and Arora, R.K. 1978. Wild edible plants of India, in *ICAR*. New Delhi.

Singh, V.K and Abrar M. Khan. 1989. *Medicinal plants and folklores: A strategy towards conquest of human ailments*. New-Delhi: Today and Tomorrow's Printer & Publishers. (A Glimpses in plant research. Vol.9).

Singh, K Anurudh. 2010.Probable Agricultural Biodiversity Heritage Sites in India: V. The Garo, Khasi, and Jaintia Hills Region. *Asian Agri-History*. Vol.14(2), 133-156.

Singh, Senjam Jinus; V.K Batra, Sanjive Kumar Singh and Thiyam Jefferson Singh.2012. Diversity of

underutilized vegetable crops species in North-East India with special reference to Manipur: A review, in *NeBIO*. Vol. 3(2), June, 87-95. *www.nebio.in, accessed on 02-03-2014.*

Yuhlung, Cheithou Charles and P.G Jangamlung. 2016. *Northeast India Tribal Studies: An Insider's View*. New Delhi: Astral International Publication.

Yuhlung, Cheithou Charles and Mini Bhattacharyya. 2016. Indigenous medicine used by the Maring tribe of Manipur, Northeast India. *Journal of Ayurvedic and Herbal Medicine*. 2(4): 146-153.

Yuhlung, Cheithou Charles. 2018. *Ethnomedicinal plants of the Maring tribe of Manipur, Northeast India*. New Delhi: Regency Publication.

Yumnam, Rajesh Singh; CH. Onita Devi, Santosh Kumar Singh Abujamand D. Chetia.2012. Study on the Ethnomedicinal System of Manipur. *International Journal of Pharmaceutical & Biological Archive*, 3(3), 587-591. www.ijpba.info http://www.manipurforest.hadrontechs.com/forest/timbers-of-manipur.html

https://www.flowersofindia.net/catalog/slides/Lomba.html

https://en.wikipedia.org/wiki/Elsholtzia

Chothe Respondent Photographs

H. Thambaljao

Y. Tomalsing (Late)

M. Merachao

M. Wounchung (L.) and Kh. Khedon (Late)

Kh. Hamunchao

Mk. Neilut (Satrapa)

M. Kundo

Y. Leingam (L.)

Th. Sirochong

Y. Maipak

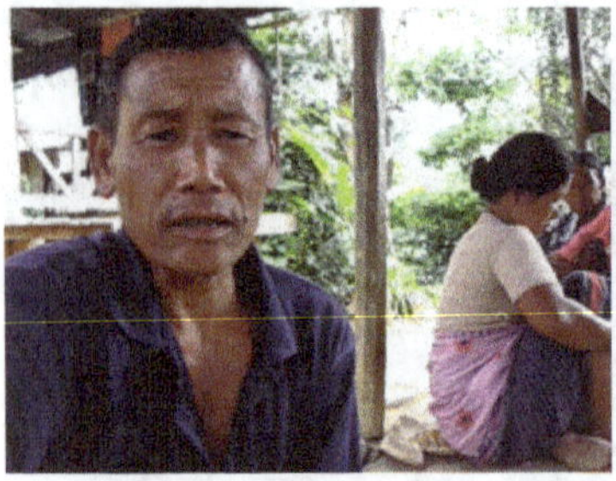

Mk. Chondin

H. Ibochou

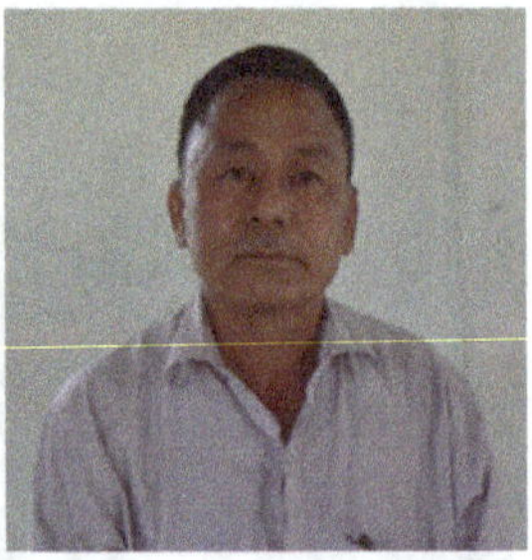

H. Mangichao

Pr. Pouti, Y. Damsu and Pr. Roushi

Pr. Pouti and Pr. Roushi

H. Joy

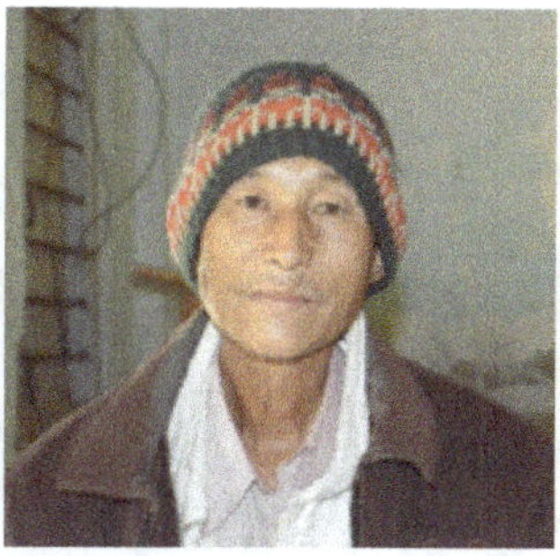
P. Bidam

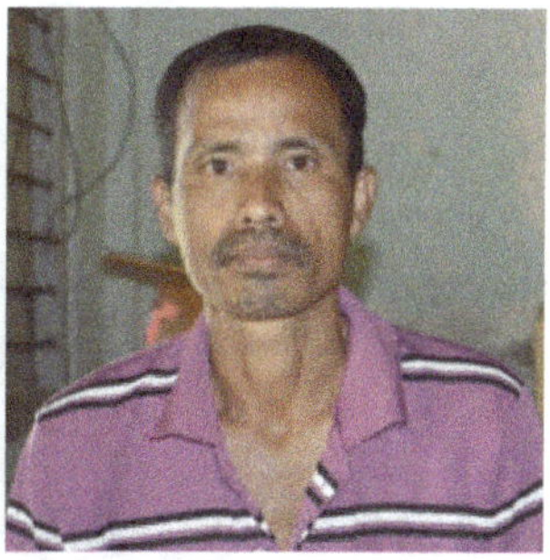
H. Cheidon

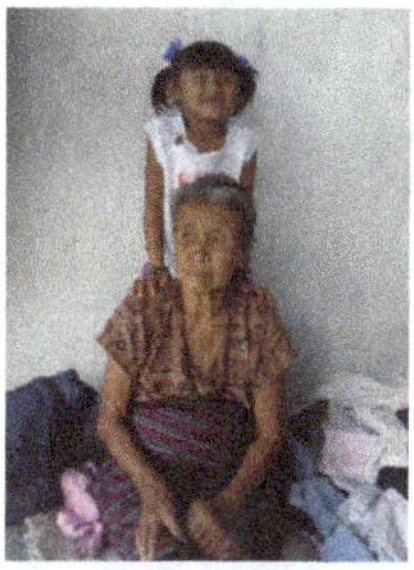
Kh. Tharaklei (Late)

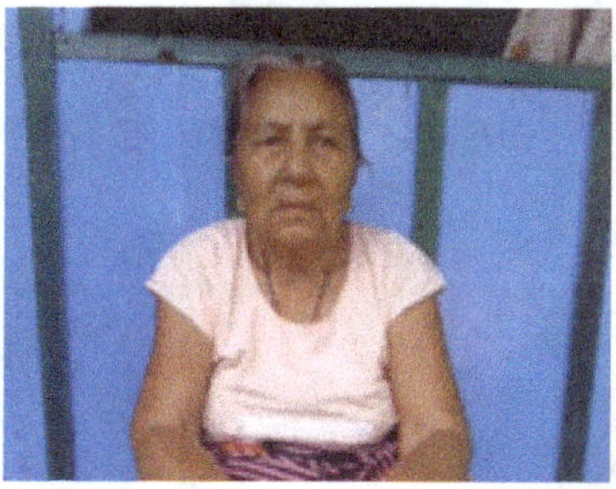
Y. Atonsicha

M. Punsi

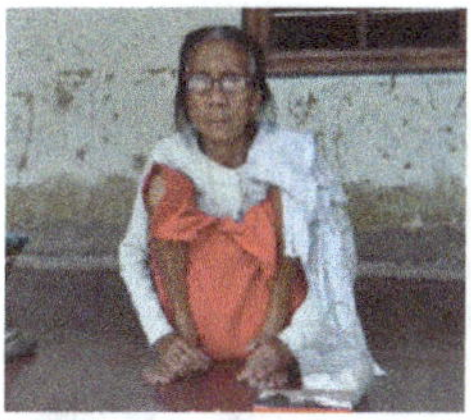
H. Punuleima

H. Jaboti

Mk. Lungnem

Y. Komani

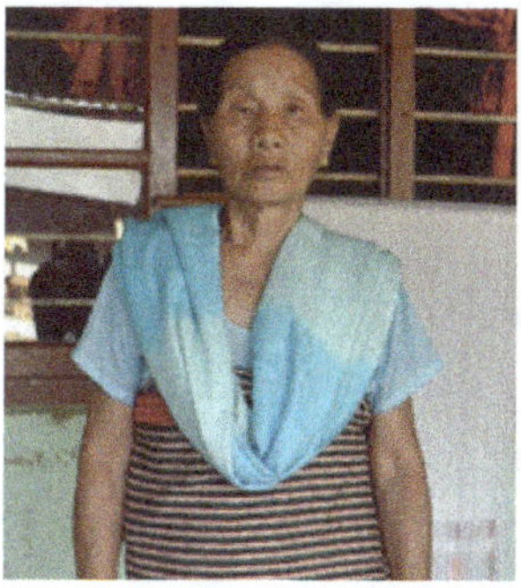
Y. Amaru

P. Akpinu (Mary)

Y. Nalani

Th. Agollu

Y. Anoupi

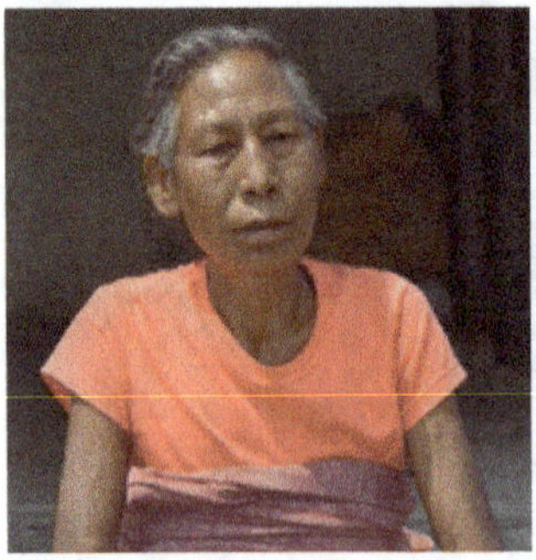

H. Atonpi

Th. Yumchaoleima

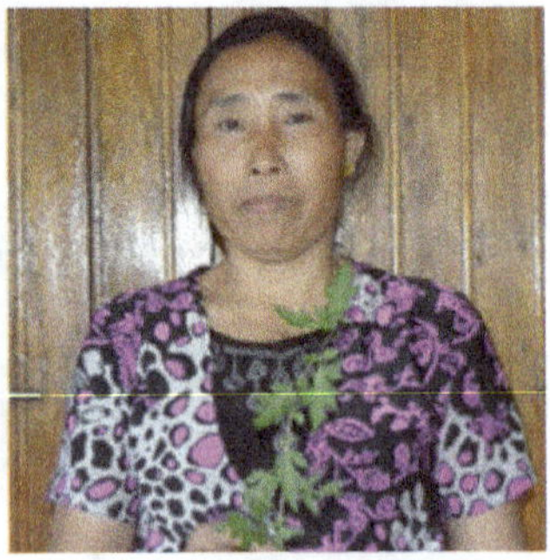

Pg. Lotamani

Ms. Borni

M. Shangkurei

H. Hoinu

H. Asin

H. Hoirei

Index

D

E

F

G

H

I

J

K

L

R

S

www.ingramcontent.com/pod-product-compliance
Lightning Source LLC
Chambersburg PA
CBHW062057160726
48282CB00003B/27